Volume II
Ciarcia's Circuit Cellar

by Steve Ciarcia

BYTE/McGraw-Hill
70 Main St.
Peterborough, N.H. 03458

Library of Congress Cataloging in Publication Data

Ciarcia, Steve.
     Ciarcia's Circuit cellar.

     Articles written by the author for a Byte magazine column, Ciarcia's Circuit cellar, which began in Nov. 1977.
     Includes bibliographies and index.
     1. Minicomputers—Collected works.   I. Title.
II. Title: Circuit cellar.
TK7888.3.C58   621.3819'5   78-20920
ISBN 0-931718-07-4 (v. 1)

678 HDHD 876

Cover Photo by Charles Freiberg

For Virginia, Chris, and Gordon
Their teamwork, strength, and dedication
is an inspiration for all.

Table of Contents

INTRODUCTION

Ciarcia's Circuit Cellar, Volume II is a collection of all my articles published in *BYTE* magazine between December 1978 and June 1980. Like many authors, I go through "phases," or stages of intense interest in specific subject areas. During that period I was specifically concerned with home control. Most of the nineteen articles focus on computer usage in the home. Subjects range from alarm systems and AC remote control to the computer control of a wood-burning stove.

The fact that I have any phases at all is significant. My relationship with *BYTE* fortunately allows me great freedom in selecting subjects, and my monthly articles are a reflection of my current interests. Anyone who writes extensively will envy me this. Of course, I'd rather have the publisher think I sweat bullets getting out an article, but the fact is that the circuits I design and prototype each month would be produced anyway. The articles are really just a method of documenting the results.

My "home control" period started in late 1978. All the depressing news stories about the rise in crime were getting to me. Although none of the houses in my neighborhood had been hit, I felt it was only a matter of time. I considered the usual alarm company installations but dismissed them as mediocre. During their first time through the slammer, most burglars learn how to get through the simple stuff. I didn't merely want to detect a crime in progress, but if possible prevent it. Failing that, if any fellow was dumb enough to break into my house with all the warnings I'd given him, he'd deserve a purple heart when the police arrived.

Obviously, only by using a computer could I design a sophisticated security system that could not be beaten. In all honesty, I may have gotten a little carried away. I am a process-control (electronic) engineer, and I've designed automated control systems for factories. Consequently, the concept of closed-loop computerized control is not alien to me; I simply visualize my house as a tomato canning factory that should be as secure as Fort Knox. The results of this pursuit are documented in a series of articles on how I applied a computer to control my home environment and security.

The first of the applications is outlined in the three-part series "Computer-Controlled Security System," in which I present the basic elements of a computerized security system and discuss similarities to computerized process-control situations.

In the construction of the security system, the input/output section receives the greatest emphasis. Schematics are given for a variety of analog and digital sensors to detect such calamities as power failure, low temperature, ultrasonic and infrared intrusion, and brownout. Control outputs range from simple contact closures and solid-state relays to hard-wired AC remote-controlled lighting and appliances.

During the year after I implemented this system, I also documented many enhancements. A "Real-Time Clock" facilitates more precise control of time-activated events. The clock circuit is made from inexpensive components and can be attached to any computer.

Similar reasoning prompted the design of a "Self-Refreshing LED Graphics Display." With so many sensors and control outputs in operation, this 8-by-16-LED array serves as an annunciator panel that instantly gives the status of the entire control and security system. It is necessary to have a display to keep track of what is going on. Without interrogating the computer program, it is possible to sit in the Circuit Cellar and watch someone approach the house, walk on the deck, and come through the door.

The culmination of the computer-controlled home concept comes in a pair of articles. With "AC Remote Control," computer control becomes a domestic reality. Rather than the hard-wired AC system (at approximately $200 per outlet) that earlier designs required, I adapt an inexpensive ($15 per outlet) carrier-current control system from BSR (USA) Ltd to allow any computer to control lights and appliances within the home. In the original design, I used hard-wired AC remote control. This entailed running a wire directly from the controlled outlet to the breaker box where a latching relay was installed. The relay was either manually or automatically controlled. The expense of a single (I installed 12) hard-wired outlet was more than a complete BSR system with 10 receivers. This is one occasion when being ahead of the times might not have been a good idea. You, fortunately, now have both options.

With the second article, "A Computer-Controlled Wood Stove," many readers probably thought I had stumbled into the Twilight Zone, that is, until they realized that the stove in question was a hydronic wood stove that plumbs hot water directly into the central heating system. The computer monitors and distributes the heat by controlling pumps and valves. Although computerization isn't necessary, a stove like this has complicated controls that require constant monitoring to alleviate potentially dangerous situations. Besides handling obvious activities, the computer provides long-term recording and analysis of data.

My "home control" phase ended abruptly when I ran out of things to computerize. Actually, it was difficult to justify more expansion within my house. Even my two Scottish terriers were hesitant to get involved. I think they may have read my thoughts when I once considered constructing a treadmill-driven generator as an emergency power source for my computer (two dog power).

I even tried wiring myself in "Biofeedback Control," but the time had come to look outside the Circuit Cellar at the rest of the computing world. I started by investigating the Intel 8086 and 8088 and then presented a variety of communication interfaces for the TRS-80. Such projects as the COMM-80 serial/parallel I/O interface introduce the reader to computer timesharing.

I have come to realize that reading my articles in book form is perhaps the best way. When I investigate a subject and write about it, my interest does not terminate with the article. My involvement continues. As my security and control system demonstrates, I develop other ideas and concepts that often enhance the material in previous articles. Each, of course, is an independent project, but they are often related by subject. Since the projects are presented together, an experimenter can read the initial presentation together with my later thoughts and determine a combination which best suits his or her needs.

There is no way for me to know how many readers actually build all my projects. From all the mail I get, a few are apparently trying. However, beware: There is a form of computer insanity that goes with living in a cellar with the only illumination coming from a CRT and the pilot light on a soldering gun. The first symptom is an incredible desire to design ridiculous gadgets. The second is telling people about them.

Steve Ciarcia
December 1980

Volume II
Ciarcia's Circuit Cellar

Photo 1: Hewlett-Packard 5082-7340 hexadecimal character display, which uses a pseudo 7 segment dot matrix. On and off control of the dots at the end of each segment allows the circuit to display capital Bs and Ds. The display pictured is powered.

Build an Octal/Hexadecimal Output Display

"Steve, I think we have a little problem!" Ray charged into the basement and hovered over me waiting for a response.

I slowly rotated in my swivel chair. The rate was barely sufficient to overcome static friction, but I finally made it. As I raised my head to talk I was interrupted.

"Steve, I think we have a problem with that EROM." Before he could finish, his expression abruptly changed and almost without a pause he ended the sentence with, ". . .what happened to you? You look like death warmed over!"

I could barely see the person standing before me with his hands on his hips. I also experienced a strange sensation of either a veil covering my face or an advanced case of furry eyeballs. Whatever the cause, Ray was still standing there awaiting a reply. It was a chore to speak. As the muscles contracted to produce the necessary air flow, I could sense a sudden recurrence of physical problems which I had hoped were on the wane.

"Steve you look terrible! You should be raring to go after two weeks in Acapulco, basking in the sun."

Ray was referring to an engineering consulting job I had just completed in Acapulco for CBS. The Miss Universe Pageant, which was broadcast live from Mexico, included a new twist this year. A computerized judging system. It sounded like a fun consulting job as opposed to the usual, "design me a computer for . . ." type. The final rationalization was, I needed a vacation anyway. I wouldn't want anyone to think that the 70 contestants had anything to do with my decision to go.

The other lucky members of our engineering party were Gus Calabrese (formerly with Digital Group) and George Watson and Dale Walker of CBS. Gus and I maintained the hardware; Dale supported the software; and, while George's official function was the electrical scoring system, his unofficial title was chief taco tester. He had this uncanny ability to sort through all the various smells emanating from a restaurant and evaluate palatability. If he didn't turn green as he walked through the door, it was Amercianized enough for us to eat there.

This smooth sailing trip was punctuated by a succession of daily crises. For instance, George's wife, having thoughtfully packed his suitcase without underwear, gave us the hoped-for opportunity to take a crash course in Mexican capitalism and to venture out to the market place. The cab driver who "drove" us there (I use the word loosely) was subject to suicidal fits. From then on everything went downhill. The list goes on and on. Reliving the past two weeks in my thoughts heightened the sense of physical malaise I was experiencing. Fortunately, Ray spoke again in time to bring me back to reality.

"What's wrong with you?"

"Let's just say it has something to do with a guy called Montezuma."

"You're not supposed to drink. . . ."

"Yeah, I know! Don't drink the water!"

Ray looked at me and decided his problem still needed attention, even though I was dying. "Steve, I was about to check the EROM contents against the listing you gave me when I noticed that it was in octal. We need to use that EROM tomorrow and we had better find the error in it tonight. I made a hexadecimal dump of the EROM contents but I still can't check it against your listing."

The response was obvious. "Why don't you convert it by hand?"

"Sure," said Ray, "I can convert it, but a thousand conversions is more than I have time for tonight. Can we assemble it in hexadecimal on your system?"

My temples were starting to throb. I hadn't used my computer in three weeks. Nothing was hooked up and I was in no condition to either attach and fire up my own programmer or write the simple algorithm to perform this minor calculation. It was hard enough for me to remember how to operate my own system without explaining the intricacies to Ray.

"Look, Ray, any night but tonight. I've got it in octal, decimal, hexadecimal, binary, —anything you want, but not tonight. I just don't think I can hack it. You understand, don't you?"

He was disappointed, but being a good friend he understood. "Can I borrow your TI programmer and some desk space? A thousand entries times five button pushes

. . . shouldn't take more than an hour or two. Got your battery charger handy?"

It seemed a shame to make Ray go to such lengths. If my system were up it would take only a matter of seconds to print out Ray's listing. It may have been a very powerful Z-80 computer on any other occasion but tonight it wasn't processing anything.

As I reached for the calculator in my briefcase I spied a relic that might provide a solution to the problem. "Ray, see that rectangular box with all the printed circuit boards plugged into the top of it?" I pointed to a bookcase that contained everything but books. "Bring it here and plug it in, and search through that pile of tapes over there until you find one marked with the same name as your listing. I made a binary dump on tape at the same time I made your listing." There are some advantages to being ill—letting others fetch and carry is one of them.

Relying mostly on Ray's high level of hardware expertise, interspersed with whatever limited verbal input I could manage, we successfully fired up my Scelbi-8B 8008 microcomputer. Even though I hadn't used it for well over a year, the read-only memory based operating system brought it to life immediately. The recognizable pattern on the light emitting diode (LED) display indicated it was ready to read input data, so I slapped in the cassette that Ray had found. Fortunately the data was stored in a format acceptable by both machines, and totally independent of the processor. I couldn't execute the Z-80 EROM listing I had loaded, but I could display it.

"OK, Ray. Now that we've loaded the data we can step through it on single step and look at it on this output port display, which I built a while back."

"How's that going to help?" Ray looked at the 3 character display as he pressed the single step a few times. "The 8008 is an octal machine. Even the data on your display is coming out octal," he said.

It was hard to smile but I managed a slight variation on the theme as I said, "Flip the switch next to the display." Instantaneously, the 257 previously displayed changed to AF, its hexadecimal equivalent.

"Hey, that's not bad, a combination octal and hexadecimal display! All I have to do is step through and copy down the hexadecimal equivalents, right?"

I nodded and Ray started to write. Barely ten entries had been made when his hardware curiosity got the best of him. "I was thinking of putting one of these on my system but it looked like too many components. By the way, I only see

Input Code				82S23 Program								7 Segment Display
D	C	B	A	d7	d6	d5	d4	d3	d2	d1	d0	
0	0	0	0	0	1	1	1	0	1	1	1	0
0	0	0	1	0	1	0	0	0	0	0	1	1
0	0	1	0	0	1	1	0	1	1	1	0	2
0	0	1	1	0	1	1	0	1	0	1	1	3
0	1	0	0	0	1	0	1	1	0	0	1	4
0	1	0	1	0	0	1	1	1	0	1	1	5
0	1	1	0	0	0	1	1	1	1	1	1	6
0	1	1	1	0	1	1	0	0	0	0	1	7
1	0	0	0	0	1	1	1	1	1	1	1	8
1	0	0	1	0	1	1	1	1	0	0	1	9
1	0	1	0	0	1	1	1	1	1	0	1	A
1	0	1	1	0	0	0	1	1	1	1	1	b
1	1	0	0	0	0	1	1	0	1	1	0	C
1	1	0	1	0	1	0	0	1	1	1	1	d
1	1	1	0	0	0	1	1	1	1	1	0	E
1	1	1	1	0	0	1	1	1	1	0	0	F

Table 1: Program for IC2 in figure 1.

two chips. Where are you hiding the rest?"

"Remind me to tell you when I recover."

Build a Combination Octal/Hexadecimal Display

Some people may consider hexadecimal displays a trivial addition to an expensive computer system, but sometimes these little add-ons make program debugging easier. I can't help but wonder whether other computer experimenters would have need for such a display. I don't expect it to replace the video display; but often, when debugging a program, it's nice to be able to display a byte here and there to verify proper program execution. It will never replace the stepper and breakpoint monitor I now use, but it's great to display keyboard or IO data quickly with a single output instruction.

There are many methods to display hexadecimal numbers on a 7 segment LED. Figure 1 and table 1 show an example of the usual brute force method using a read-only memory as a hexadecimal decoder. Programming the 82S23 was described in the November 1975 BYTE ("A Versatile Read Only Memory Programmer," by Peter H Helmers, page 66). While this is a viable approach, an excessive number of components is needed in this stand alone display, and most people would rather not have to program an EROM. The alternatives are to allow a computer to perform the decoding and drive the 7 segment display through the transistors from a latched 8 bit output port, or to put additional logic around a standard 7 segment decoder driver for the extra requirements. The former case necessitates a computer program while the latter can involve as many components as those in figure 1.

Fortunately there are other products on the market that can solve the problem. I've been using the Hewlett-Packard HP7340 hexadecimal display for a number of years. Those familiar with it can rightfully say how trivial the solution was, while those who are not may find it a revelation. Photo 1 illustrates the physical size of the HP7340. A hexadecimal A is displayed in a dot pattern. These hexadecimal displays depart from standard 7 segment format by being capable of displaying a capital B and D in hexadecimal. This is accomplished by controlling the corner dots which give the appearance of "rounding." This ability discriminates a B from an 8 or a D from a 0. There are 16 distinctly different characters.

An additional feature of the HP7340 is that it contains a 4 bit latch and the decoder/driver as well. The result is a single 8 pin hexadecimal display which successfully accomplishes the function of all the circuitry in figure 1. The specification of the individual pins are in figure 4.

Figures 2 and 3 demonstrate how the HP7340 can be configured to function as a 2 digit hexadecimal output port or a 3 digit octal port. No 8 bit latch is required since it

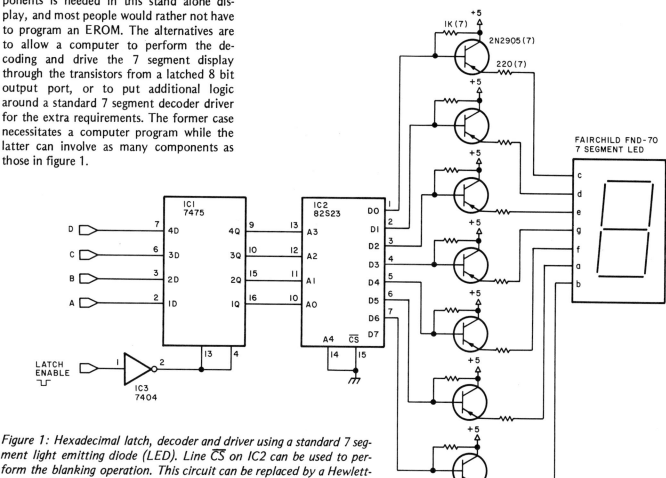

Figure 1: Hexadecimal latch, decoder and driver using a standard 7 segment light emitting diode (LED). Line $\overline{CS}$ on IC2 can be used to perform the blanking operation. This circuit can be replaced by a Hewlett-Packard HP 7340 or equivalent (see table 2).

Photo 2: Prototype board of the circuit in figure 4. Two similar circuits were built on the same board. When in the hexadecimal mode (shown at left in the picture), the leading digit is blanked. The display at the right shows the octal mode. Each is wired as an independent output port, but the computer sends the same data to both.

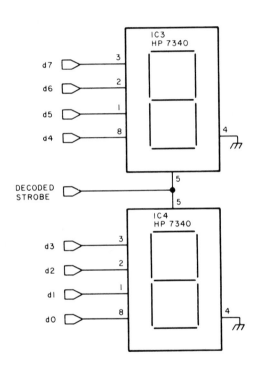

Figure 2: Hexadecimal latch, decoder and driver display circuit.

Pin Number	Function
1	Input B
2	Input C
3	Input D
4	Blank Control (blank = +5 V)
5	Latch enable (latch = 0 V)
6	Ground
7	+5 V
8	Input A

Table 2: Pin functions for the Hewlett-Packard HP7340 binary coded decimal (BCD) to hexadecimal display. Similar displays are made by Dialite and Texas Instruments.

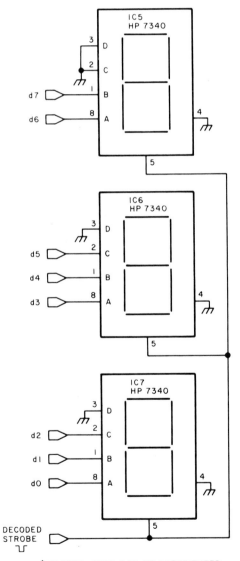

*HP 5082-7300 CAN BE SUBSTITUTED FOR HP 5082-7340 IN OCTAL READOUT APPLICATION. 7300 IS NUMERIC ONLY.

Figure 3: Octal latch, decoder and driver display circuit.

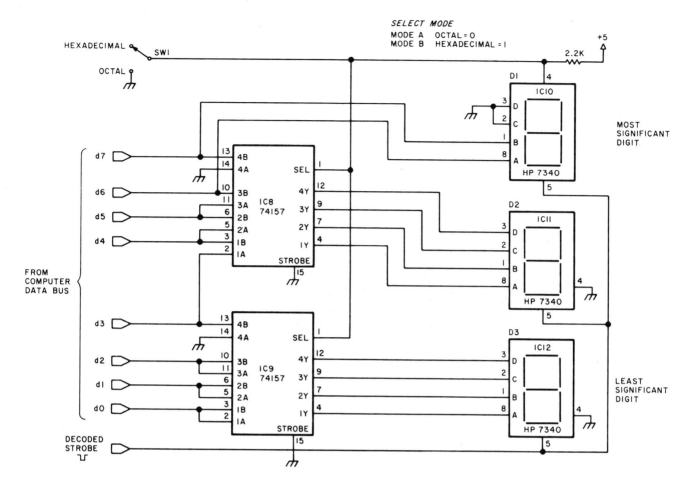

Figure 4: Combination hexadecimal and octal display circuit.

Number	Type	+5 V	Gnd
IC1	7475	5	12
IC2	82S23	16	8
IC3	HP7340	7	6
IC4	HP7340	7	6
IC5	HP7340	7	6
IC6	HP7340	7	6
IC7	HP7340	7	6
IC8	74157	16	8
IC9	74157	16	8
IC10	HP7340	7	6
IC11	HP7340	7	6
IC12	HP7340	7	6

Table 3: Power wiring table for figures 1, 2, 3 and 4.

already contains one. The 7340s can simply be attached to the data bus at any other parallel output port and strobed from a chip select decoder.

Figure 4 is the circuit of the unit similar to the one Ray used. Two multiplexer circuits alternate the input connections to the displays so that when switch 1 (SW1) is in the octal position, the circuit performs as figure 2, and when in the hexadecimal position, as figure 3. The leading character is blanked when in the hexadecimal mode. Two of these circuits are combined in the prototype board of photo 2. The left display is in the hexadecimal mode showing B7 while the right is in the octal mode display-ing an equivalent 267 octal. The same binary information is being sent to each port; only the switch setting differs.

Usually these or equivalent displays are advertised only as hexadecimal displays. All strictly hexadecimal displays that I've seen contain these same electronics. While alphanumeric displays will also work, they require extensive scanning logic and are an overkill for this application.

In Conclusion

I hope this simple circuit will eliminate any frustration you may have in the area of hexadecimal displays.

One question I'm often asked is whether my introductions are true. So far everything I've written is based upon actual people or events. While I take considerable poetic license in describing the situations, it is not necessary to invent fiction when experience is often so much more humorous.■

Build a Computer Controlled Security System for Your Home

Software consulting by Steve Sunderland

Ciarcia's Circuit Cellar

Part 1

"Steve? Lloyd? How did you get in?" I jumped back in my swivel chair in surprise as they approached.

"Joyce let us in."

It wasn't unusual for Steve and Lloyd to drop over. We've been friends and fellow computer buffs for years. The surprising part was that they hadn't seen my new basement since I had moved and they decided not to wait for an invitation.

"Boy, when you said you moved to the woods, you really meant it." I was sure that a western Long Islander like Lloyd probably considered more than three trees and a hedge in a yard a forest, but I did have to admit that the last 100 yards along my dirt road was a real killer. "How do you plan to get out of here in the winter?"

A picture of Nanook of the North complete with dogsled and team flashed into my mind. I felt the cold wind blowing in my face and frosty icicles forming in my bushy beard. Protecting my eyes from the blizzard conditions with a heavily gloved hand, I searched the horizon for the faint wisp of smoke rising from a chimney that would provide some respite from the harsh winter. The whip in my other hand was glazed with ice and no longer produced the crisp snap that was necessary to command the attention of the lead sled dog. The iciness enveloped me and began to cloud my consciousness. I could only hope that instinctual survival reactions would be triggered in time. . . .

"Steve! What are you going to do in the winter?" Lloyd repeated, breaking me from my trance.

Thoughts of my arctic ordeal instantly dissolved and the true answer came to mind. "A Jeep came with the house. Come over when it snows and I'll show you how to plow a road."

Lloyd shuddered. The thought of being bounced around in a Jeep while plowing a rutted dirt road was more than he could take. "No thanks. Maybe I'll watch — from inside!"

Before I could discuss the merits of mechanized dogsleds and their relationship to my fantasy, Steve asked a question I had been concerned about but had not yet resolved. "Have you thought of a security system for this place?"

"Actually, I've done more than think about it, I've started to design one." I was aware of my new isolation, and while crime was an important consideration, any security system I designed would do far more than just ring an alarm bell.

"Yeah! You ought to see the system I have," Lloyd piped in. "It cost a fortune, but anybody breaking a window or opening a door will set it off. It automatically dials the police too."

"Lloyd, I've been to your house. It's wired like Fort Knox! You've got foil tapes on every window like a jewelry store. This house is a contemporary, if you didn't notice as you entered. Do you realize how much glass there is in the living room alone?"

"It's a pretty foolproof method," he answered as though not hearing my question.

"Well . . . ," I continued, "there are over 300 square feet of glass in that one room. All I need is 300 square feet of stripes. And didn't you say you often had false alarms in the spring and fall when the temperature variations induced cracks in the tapes?"

"Hold it Steve. Lloyd was only talking about one possibility. You don't have to install the same system. In fact there must be dozens of commercial burglar alarms you could have installed."

Steve was right. "I'm sorry guys. Moving out here in the woods has pointed out the need for self-sufficiency and independence. That's why I have the wood stove over there. I'm even going to install an auxiliary generator to power the water pump and lights. The only item I haven't settled yet is the security system."

"Isn't it a simple matter of stringing some door and window switches, an on/off key switch, and an automatic dialer?"

"Sure Steve, if I wanted a simple burglar alarm. You should know me better than that, though. Reasonably priced commercial systems are nothing more than what you've described. As soon as you add time related activations, battery backup, intrusion sensing to cover approach roads and surrounding property, and a decision making capacity from the alarm controller, dedicated logic becomes too expensive."

"What kind of system do you have in mind?" inquired Lloyd.

"Actually I haven't finalized all the details, but the system I want would do things like turn on the outside lights as I drive up to the house, activate 115 VAC appliances in either a preprogrammed sequence or randomly, have display panels in the bedroom and down here in the basement to track

anyone approaching the house, and of course all the usual fire and burglar alarm functions."

"Wow, Steve!" Lloyd's eyebrows rose a little as the security system was unveiled. "That would take a computer!"

"Exactly!" I exclaimed.

Steve looked at me, then at the big (by microcomputer standards) computer to my left. His technical mind did a quick calculation on the battery backup requirements for my 64 K dual disk computer system.

"I suppose you've laid in a separate power feeder from the Grand Cooley Dam to keep this dinosaur running?" he said.

"Don't be funny. The computer I propose to use is a single board microcomputer, programmed to provide the sophistication I want but versatile enough to allow logic changes and easy hardware expansion." I reached into the drawer behind me and pulled out a foot square populated printed circuit board and passed it to Steve. "Here's the computer." Both Steve and Lloyd eyed the odd looking

SECURITY SYSTEM OUTPUTS

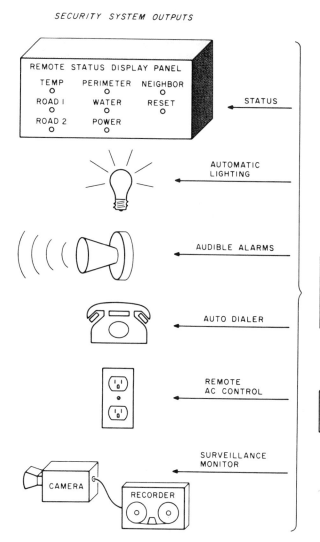

Figure 1: Pictorial diagram of computer controlled home security system.

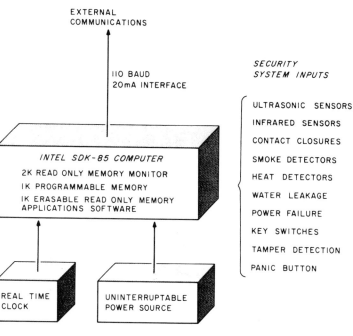

EXTERNAL COMMUNICATIONS

110 BAUD 20mA INTERFACE

INTEL SDK-85 COMPUTER

2K READ ONLY MEMORY MONITOR
IK PROGRAMMABLE MEMORY
IK ERASABLE READ ONLY MEMORY APPLICATIONS SOFTWARE

REAL TIME CLOCK

UNINTERRUPTABLE POWER SOURCE

SECURITY SYSTEM INPUTS

ULTRASONIC SENSORS
INFRARED SENSORS
CONTACT CLOSURES
SMOKE DETECTORS
HEAT DETECTORS
WATER LEAKAGE
POWER FAILURE
KEY SWITCHES
TAMPER DETECTION
PANIC BUTTON

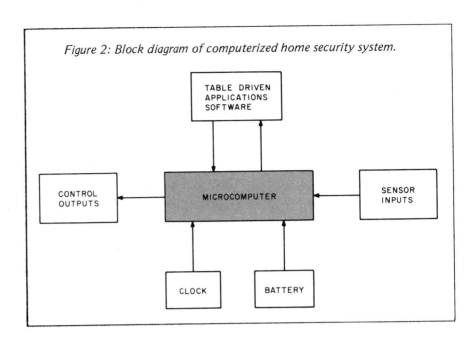

Figure 2: Block diagram of computerized home security system.

TABLE DRIVEN
APPLICATIONS
SOFTWARE

CONTROL
OUTPUTS

MICROCOMPUTER

SENSOR
INPUTS

CLOCK

BATTERY

circuit. "It's an Intel SDK-85 kit which uses an 8085 processor. With a little additional programmable read-only memory to store the program, some ultrasonic and infrared sensors, contact switches at strategic locations, a house wired for remote AC control of lights and appliances, and about a mile of wire, I'll be in business. I only pray for divine inspiration as I sit down to assemble the program."

"Hey, this sounds fascinating. Why *not* a computer controlled security system!" Steve seemed more elated at the prospects of my design than I did. He was a control systems engineer by trade; I could sense he was already writing the control algorithms in his mind.

"You wouldn't want to help me write the software would you?" I asked.

"Sure, sounds like fun."

"Only you would consider writing software fun." Lloyd pointedly remarked to Steve. "I'm more interested in the hardware." Turning to me he continued. "Tell me about your design. Is the SDK-85 inexpensive enough to dedicate to an alarm system? What kind of a battery backup supply are you making? What kind of sensors are you putting outside on the roads? Did you add that infrared scanner you already wrote about?"

"Hold it guys. One question at a time. Maybe I'd better start from the beginning."

Anatomy of a Computerized
Security System

The primary requirement for the hardware and software used to implement a sophisticated home security system is flexibility. Considerable flexibility is required to allow you to structure a system which executes a variety of functions. These functions include not only the menial tasks of notifying residents that an intruder has entered the home and signaling the authorities, but also those activities necessary to enhance one's "quality of life." Such items include starting the electric percolator at the proper time each morning or turning on the cooling system in the wine cellar when the temperature goes above 64°F.

The system to be introduced in this article and detailed in Parts 2 and 3 has the hardware and software features of a sophisticated home control and security system, a system which may also be adapted to perform some of the basic routine control functions you may want to perform around the home or office (see figures 1 and 2). It has three major components:

1) The microcomputer and its associated hardware.
2) Event-table-driven software set.
3) A sophisticated array of sensors.

Each of these items will be detailed in the succeeding sections of this series. Part 2 will describe the Intel SDK-85 single board computer which satisfies the intelligence requirements of the security system, the hardware modifications necessary to add additional programmable and erasable programmable memory to the SDK-85 kit, and a description of the various control modules comprising the software set of the system. The third article of the series will discuss sensor design, uninterruptable power supply requirements, design and construction of an indicator panel to display monitored conditions, and, finally, example software that

Photo 1: The Intel SDK-85 single board computer, that forms the heart of the computer controlled security system described in this article. Note the additional erasable read-only memory and programmable memory added in the prototyping section at upper left. For more information about the SDK-85, contact Intel, 3065 Bowers Av, Santa Clara CA 95051

demonstrates the versatility and ultimate capability of a computer controlled security system.

By this time you may be wondering why I used an Intel SDK-85 microcomputer in building this system. Before I answer this question, let's discuss the general requirements.

A computer home security system's link with the real world is its sensors. The sensing devices must be small, reliable, low power, and capable of detecting exceptional sound, light, and motion. No single device, I used was capable of detecting changes in these three conditions, with the exception of the infrared scanner described in a previous article (see "I've Got You in My Scanner! A Computer Controlled Stepper Motor Light Scanner," by Steve Ciarcia, November 1978 BYTE, page 76). A variety of single function sensors provided the necessary inputs: ultrasonic devices to sense motion within the monitored area, pressure switches to monitor the removal or the addition of a mass (such as a foot on the stairs to the Circuit Cellar), photoelectric devices to detect movement between monitored points, and an array of contact closures to signal the opening and closing of strategic doors and windows.

All inputs to the computer are designed to be discrete in nature. That is, they exist as switched outputs and appear to the computer as either a high or low logic level in the activated state, depending on the particular sensor. The majority of the sensors have relay outputs and can be wired in series or parallel combinations to cover wide areas or to give redundant indication. An illustrative example is using the series combination of an ultrasonic and infrared sensor to cover large open areas. The failure or false indication of a single input will not cause a false alarm, since both sensors must have a positive "intruder" presence. In effect, the signals from the two devices are logically ANDed. More on these techniques when we discuss sensors.

The Alarm Contains More Than a Switch and a Bell

Functions other than the detection of intrusion have been incorporated in the system. These functions are centered on the safe operation of the home or office, and are used to detect the status of sump pumps, refrigeration systems and water pumps. Failures of these devices are detected by the system again via simple contact closures. The detection of smoke and fire has been incorporated into the system for the protection of the property and its occupants.

Figure 3 is the floor plan of a contem-

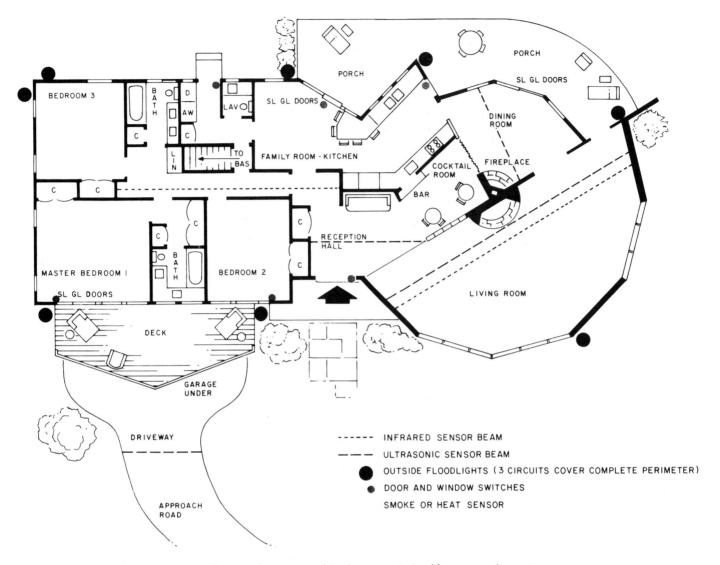

Figure 3: Typical arrangement and types of sensors used in the computerized home security system.

porary house similar to my own. For obvious reasons I have chosen not to use one of my own home. The dimensions and layout are similar enough to adequately represent the model and to illustrate the possible quantity, type and placement of sensors. I consider my house a unique application and do not believe that every detail of this system is necessary for adequate security coverage. Less sophisticated photoelectric and ultrasonic sensor designs could be used to monitor a 9 by 12 foot room, but might be inadequate in a much larger room. My home proved to be a sensor development challenge, since the living room is 42 feet in diameter and the house totals 4800 square feet monitored area. The best procedure to follow is to understand the theory of the system, differentiate between the uniqueness of my application and a general case, and extract those components which satisfy your security needs.

Alarm Priorities

The inputs monitored by this security system are divided into four priority levels, level 0 through level 3:

Level 0
Activation
Level 0 alarms are always enabled as long as the system is operational. Inputs of this type consist of smoke sensors, heat activated switches, and panic button inputs.
Result
Positive closure of any sensor immediately triggers an audible alarm within the residence. Automatic dialing of police, fire and neighbor commences 45 seconds later. This delay allows time to reset system in cases of false alarms, such as burned toast.

Level 1

Activation

Level 1 alarms are enabled by a keylock switch input to the computer and are primarily intrusion in nature. They consist of parallel and series combinations of ultrasonic, infrared, and contact closure type sensor inputs.

Result

Positive closure of any sensor immediately triggers an audible alarm inside and outside. The system immediately dials police, contacts the neighbors via a dedicated serial communication link between houses, and turns on surveillance recording devices.

Level 2

Activation

Provision is made to allow entry to the house at selected spots to deactivate the alarm. Entering these points starts a timer which must be reset by deactivating the level 1 alarm system.

Result

Failure to reset the system within the alloted time triggers a level 1 condition with typical level 1 response.

Level 3

Activation

Level 3 is always activated and its input sensors are displayed on a panel as well as being supplied to the computer for possible action. Sensor inputs include power failure, perimeter intrusion, freezer failure, etc.

Result

Perimeter or approach information can be utilized by the system to turn on outside lights, but the primary purpose of level 3 is to provide a noncritical condition monitoring system for information. The duration of the displayed event is a timed function.

The combination of the four levels uses a total of 20 parallel input bits on the SDK-85.

Alarm Outputs

Unlike the majority of security alarms, which have only a single output to turn on the automatic dialer and alarm bell, this system has 18 discrete output functions. They can be activated in any combination, and are described in table 1.

The particular designations for inputs and outputs are selected for my needs, and are easily redefined due to the program structure and event table software. A pictorial diagram is shown in figure 1. This will be explained in detail in part 2.

Additional Considerations and Capabilities

Another major consideration in a home security system is the provision of a continuous and reliable power source. This is necessary to sustain operation during electrical interruptions. To assure a dependable power source, the system is powered by an uninterruptable power supply. Necessary input voltage is provided, at all times, by 12 V batteries. The charge on the batteries is maintained by an automatic 115 VAC charger which will charge the batteries when their output voltage falls below a preset value. The output of the power source is used to supply voltage to critical sensors, the status display panel, and the microcomputer. The computer will of course register the occurrence of a power outage.

As stated previously, one capability of this system is the controlled activation of AC outputs triggered by input events or timer generated commands. To accomplish the latter, one needs a time base and a real time clock. The circuit in figure 4 is added to the SDK-85. The real-time clock creates a pulsed output with a frequency of 1 Hz. This pulse output signal, when attached to the interrupt

Table 1: 18 discrete functions performed by the author's computer controlled security alarm system.

Output	Function
1. Auto dialer	Automatically calls police, fire, etc.
2. Audible alarm (intrusion)	Signals that there has been a break-in.
3. Audible alarm (fire)	Signals detection of fire or smoke.
4. Level 1 enabled (light next to key switch)	Steady output indicates reset condition; flashing output indicates enabled mode.
5. Perimeter flood lights	Activated by timer or level 3 input.
6. Driveway flood lights	Timer activated at dusk.
7. Low level audible alarm	Signals pertinent combination of events or particular level 3 activation; mounted in level 3 display panel.
8. Recording instruments	Particular level 1 alarms trigger activation of surveillance recording devices.
9. 10. 11. {3 bit, level 3, display board driver}	Multiplexed output controls eight display parameters.
12. AC circuit 1 13. AC circuit 2 14. AC circuit 3 15. AC circuit 4 16. AC circuit 5 17. AC circuit 6	Six AC power control outputs to lights or appliances which either simulate occupancy through sequential activation or act as remotely controlled outputs to turn on percolators, air conditioners, etc.
18. Output to neighbor	Indicates fire or intrusion.

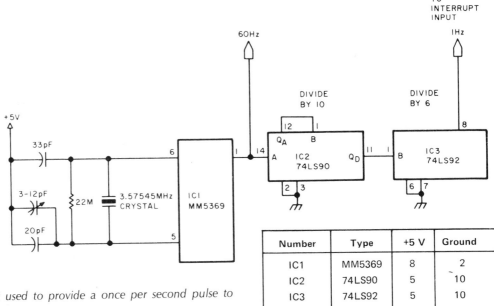

Figure 4: Real-time clock circuit used to provide a once per second pulse to update the SDK-85 time of day software.

Number	Type	+5 V	Ground
IC1	MM5369	8	2
IC2	74LS90	5	10
IC3	74LS92	5	10

input line of the 8085 and combined with proper software, provides the alarm system with a time of day clock having resolution of 1 second. The current time in hours, minutes and seconds is constantly displayed and updated on the 6 digit LED display on the SDK-85 board. This gives a constant indication of system operation and readiness. This display can be deactivated to conserve power during a power interruption.

The time of day clock is not the only operator interface. An additional display panel is remotely located in the bedroom

or similar area where current system status can be viewed. This display is not meant to indicate the obvious, but rather to make the resident aware of less critical but equally important matters. In the event of a fire or outright break-in, the system would activate the necessary counter measures and notify the authorities. Flashing a tiny light emitting diode (LED) on a panel will hardly be noticed in all the noise of the sirens.

A third level of sensors monitors approaches to the house, the perimeter, and important events such as water leaks and power failures. While the system may still turn on outside lights if desired, this panel would immediately notify the resident of someone passing through the property, or of a power failure. When you live in the woods, these are all important considerations.

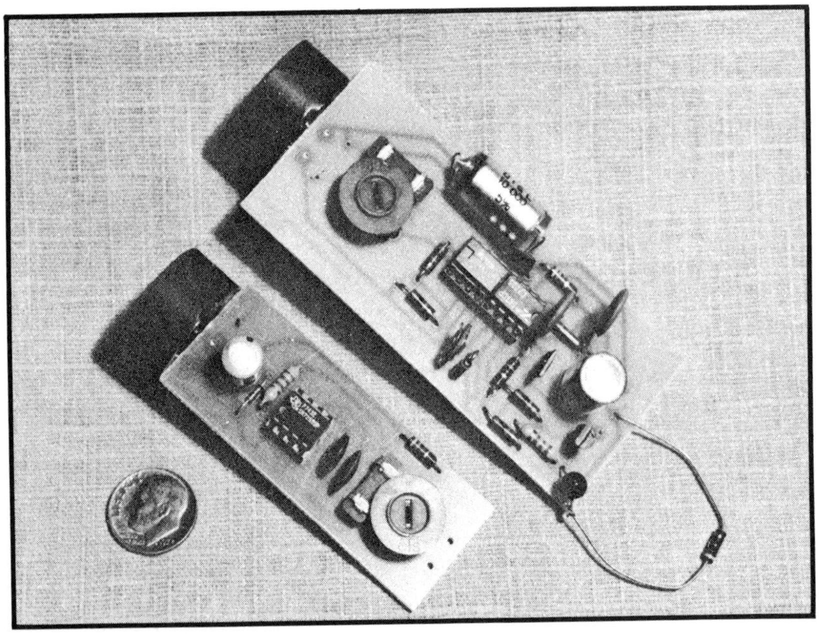

Using an Intel SDK-85

The microcomputer system utilized in this security system was constructed around an Intel MCS-85 system design kit (SDK-85). It was chosen because of the basic features it provides to the experimenter. Included in this $250 kit are all the materials necessary to build a basic computer system which can be modified easily with additional programmable read-only memory containing the security system software. The main features provided in the standard SDK-85 include:

- 8085 processor utilizing the 8080A instruction set.
- Interactive 6 digit LED display and 20 key keyboard.
- 110 baud 20 mA current loop interface.

Photo 2: Prebuilt commercial ultrasonic transmitter (small board) and receiver available from Bullet Electronics; for use across doorways, rooms, etc.

- Comprehensive read-only memory based monitor (2 K bytes).
- 256 bytes of programmable memory.
- 38 parallel input output (I/O) lines.
- Three user-vectored interrupts.

The presence of a good monitor, a quantity of I/O bits, vectored interrupt capability, and an interactive keyboard and display were the primary reasons for the choice of this unit. As previously stated, the real-time clock interrupts the system every second. While interrupt capability is not absolutely necessary to the design of a security system, it does insure that events which may become critical to the performance of a system are not missed. For example, let's assume that maintaining an accurate time base is required, and that the system maintains time by a pulsed input. If this input is interfaced to the computer via a line on an input port, it may be possible at times (depending on processor load) to miss the event, thus causing the system to slowly lose its time synchronization to the real world. However, if this input is interfaced to the system via an interrupt line, the system will not "miss ticks" due to processor overhead. Should the processor be busy at the time the interrupt occurs, two possible conditions could exist. First, the routine being executed could have temporarily inhibited interrupts. This will only delay the processing of the timer interrupt: when the interrupts are enabled by the routine the processor will recognize that the interrupt is pending and vector to the appropriate address to begin processing the interrupt. Note that, should interrupts be inhibited for a period longer than the frequency of the interrupt, "missed tick" conditions will occur. Therefore, one should always be aware of the maximum time during which interrupts will be inhibited. The second condition which can arise during the processing of interrupts is that the processor might be busy executing a module with the interrupts enabled. In this case the processor will immediately vector to the routine to process the interrupt. Upon completion of the interrupt processing, control is returned to the interrupted module.

The keyboard and display are important to the operation of the security system. These units are utilized extensively in the startup of the system.

It is through this facility that the user initiates the operation of the system, notifies the computer of the current time of day so that it may initialize the software clock, and can utilize the monitor to diagnose problems within the system. For example, the key-

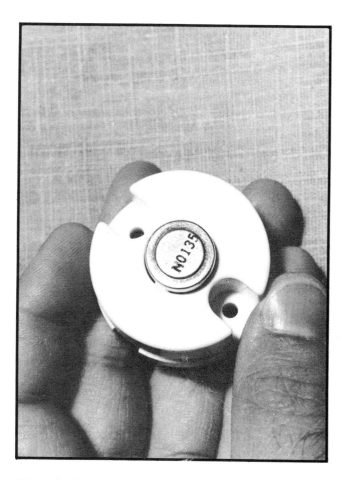

Photo 3: Heat sensor designed to trip above 135° F, typical of the type often used in combination with smoke detectors.

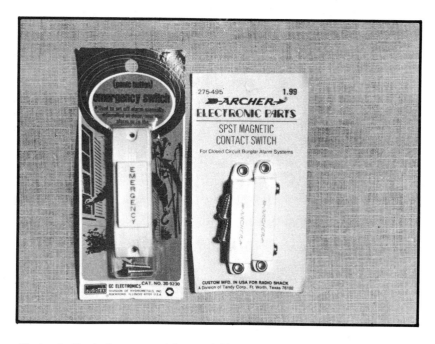

Photo 4: Typical commercially available sensors. Shown are a panic button (left) and magnetic door switch, available from GC Electronics, Rockford IL, and Radio Shack, respectively.

board and display allow the user to initialize the system by:

- Resetting the processor by depressing the Reset button.
- Entering the 4 digit address of the start of the security system cold start module, which will be displayed, as entered, on the display.
- Depressing the Go button.

The functions provided for debugging the system include:

- The ability to examine the contents of memory locations registers both randomly and sequentially.
- The ability to load new values into memory and registers both randomly and sequentially.
- The ability to single step the execution of a software module.
- The ability to generate vectored interrupts from the keyboard.

Photo 5: Steve Ciarcia installing a heat sensor over the printer and the Selectric in the circuit cellar.

- The ability to initiate execution from various addresses.

The monitor provided in the SDK-85 resides in 2 K bytes of read-only memory. It contains the software required to service the commands the processor receives from the keyboard, the routines to drive the 6 character LED display, the algorithms to process and serve Teletype (or other serial I/O device) requests and the software to handle serial I/O communication. This monitor also provides the vectored interrupt trap locations which in turn direct control to user-supplied modules.

Several of the monitor routines are very useful to the operator and were used extensively to display input data and do the initial debugging.

Four modules are provided for sending and receiving data to or from the Teletype. These are:

- Console input which returns an ASCII character received from the Teletype to the requesting module.
- Console output which transmits an ASCII character from the sending module to the Teletype.
- Carriage return/line feed which outputs to the Teletype a carriage return followed by a line feed.
- Hexadecimal number printer which converts a byte of data to two ASCII characters (hexadecimal 0 thru F) and outputs them to the Teletype.

Three modules are provided for sending and receiving data to the keyboard/LED display. They are:

- Update data which displays the contents of register A on the data field of the display in hexadecimal notation.
- Read keyboard which returns to the requesting module the character entered on the keyboard.
- Output data which outputs data to the address or data field of the display.

The security system software is comprised of a set of highly structured modules (see figure 5) used to process records passed from one module to another. Within this structure the initiating event can come from one of two sources, the detection of an external event or the occurrence of a time event. The occurrence of an external event is detected by the digital scan module; the occurrence of a time event is detected by the timer interrupt processor. When either of these modules detects the occurrence of an event that requires further processing, it passes an event index to

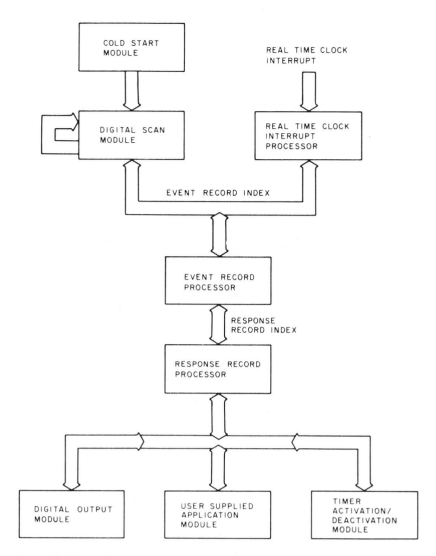

Figure 5: Block diagram of the computer controlled home security system.

the event processor, which will initiate the processing of the event. The event processor utilizes the event index to extract the response records associated with the event in question from the event record. It then transmits this data to the response processor, which will perform the required response (ie: output a contact closure to turn on a siren or autodialer, or activate a special application module). The detailed structure and functions of the various tables and modules that make up the security system software set will be described in part 2 of this 3 part series.■

Build a Computer Controlled Security System for Your Home: Part 2

Ciarcia's Circuit Cellar

"Hi, Lloyd. What brings you over this afternoon?" Lloyd often came over unannounced, so I wasn't surprised. I barely lifted my eyes from the circuit I was inspecting as he approached.

"I just wanted to see how the security system was coming. I brought along the telephone number of the alarm company that did my house, should you care to reconsider and be more conventional."

"It's more fun this way," I responded.

"I just about made it into the driveway out there. It's like Grand Central Station. What's going on?"

"You can't have remotely controlled perimeter lighting without installing floodlights." There were two electricians wiring eight sets of high intensity floodlights to a central control panel in the basement (see photo 1). This control panel in turn would be connected to the computer for automatic control of the lights. "Besides, those are high intensity 250 W lights. With three or four of them on a circuit, they had better be wired correctly."

"Electricians? Floodlights? Wouldn't it be cheaper to wire it yourself?"

"When you install this kind of system, you have to be extremely careful not to create a problem greater than the one you're trying to alleviate. Outside weatherproof cabling isn't exactly my bag. I'd much rather sit here and make the modifications to the SDK-85 controller. Sit down and I'll explain what has to be done."

SDK-85 Modifications

The program necessary for this application requires slightly under 1 K bytes of memory. Since the control algorithms are fixed and do not change, they should be written in nonvolatile storage of some kind. For our purposes an ultraviolet erasable read-only memory such as a 2708 or 2716 is recommended. As supplied, the SDK-85 contains 256 bytes of programmable memory used by the control program for stack storage and variable tables. While 256 bytes is adequate once the system is operational, additional programmable memory is suggested for checkout purposes. The larger area allows room for multiple diagnostic subroutines. Once checkout is completed, the 1 K byte memory buffer can be removed and the software readdressed to the location of the 256 byte buffer. This is not a requirement, however.

There are two ways to add additional erasable read-only memory to the SDK-85. The simplest is to buy an 8755 2 K byte integrated circuit and plug it directly into the slot already provided for extra memory on the board (this slot can accommodate an 8355 read-only memory or 8755 erasable read-only memory). In industrial applications where the latest chips on the market are no problem to obtain, this is the only reasonable approach. For the computer experimenter, however, these parts are relatively hard to find, and the second approach must be investigated.

Photo 1: View of author's house. Professional electrician Russ Molitoris is shown stringing electrical conduit pipe for high intensity outside lighting. Conduit from all external lights runs to central control panel in basement which accepts DC on/off control commands from the computer. Conduit is used to meet electrical code requirements.

The 8085 differs from the 8080 in its method of multiplexing bus information to peripheral circuits. The new exotic large scale integration circuits like the 8755 incorporate full decoding logic which minimizes external support circuit requirements. Using the 8085, a complete computer with central processor, read-only memory, programmable memory and I/O can be constructed with three integrated circuits. The addressing of these multifunction support circuits differs from the common variety of memory and I/O devices we have become familiar with. To use devices other than the 8755 and 8155 generation requires a series of demultiplexing registers and buffer drivers to break out the address and data lines to a logical equivalent to the 8080. The SDK-85 allows for optional bus expansion and provides nine blank, prewired integrated circuit locations for just this purpose.

Some of these registers and bus drivers are relatively expensive considering our requirements. With the nine integrated circuits inserted, a full 65 K bytes of expansion memory can be accommodated. The additional 1 K bytes required in the present application does not warrant this much expense. The method illustrated for expanding the memory of the SDK-85 using readily available components is predicated on its staying a small system, ie: less than 8 K bytes. The full expansion circuitry is required above this value. Table 1 shows the necessary modifications.

Photo 2 is a close-up of the header

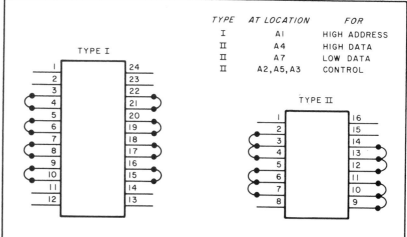

TYPE	AT LOCATION	FOR
I	A1	HIGH ADDRESS
II	A4	HIGH DATA
II	A7	LOW DATA
II	A2, A5, A3	CONTROL

TYPE I

TYPE II

Modifications to SDK-85 Board
(for small system)

Note: An Intel (single 5 V supply) 2716 can be substituted by removing the −5 and +12 V supply connections and tying pin 19 to BA10. (Numbers with J prefixes refer to connector pins on the SDK-85 board. Numbers with A prefixes are integrated circuit locations.)

- Solder in 40 pin wire wrap headers at J1 and J2.
- Solder in 34 pin wire wrap headers at J3.
- Solder in 26 pin wire wrap headers at J4 and J5. (I used Scotchflex wire wrap headers.)
- Insert 8212 at A6 to hold low address lines mutliplexed.
- Insert jumper headers as in table 1.
- Add 74S00 at A8 and 74LS74 at A9 to enable line DSI 8212 (except during HOLD). Other buses are unbuffered, and will be floated at 8085.
- Meaning of signals on J1 and J2 remains the same as with SDK-85 circuitry, just less drive level for a small system.
- Now add standard 8080 I/O memory devices via J1 and J2 wire wrap posts.

Table 1: Modifications which must be made to the SDK-85 board to expand the memory capabilities.

Photo 2: To add the additional memory to the SDK-85, nine integrated circuits are required. If a limited quantity of expansion is required, the technique described in the text using three integrated circuits and jumpers can be used. This photo illustrates these jumper headers installed in the circuit.

Software consulting for this series of articles was provided by Steve Sunderland.

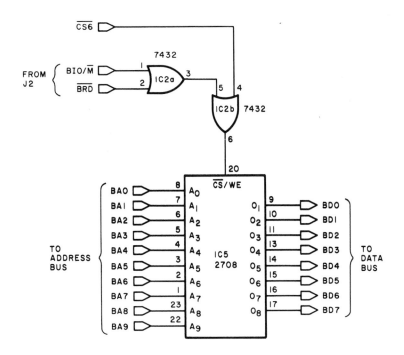

Figure 1: Circuit for adding a 2708 erasable read-only memory to the SDK-85 computer board. Note that an Intel 2716 (single 5 V supply) can be substituted by removing the −5 V and +12 V power supply connections and tying pin 19 to BA10.

Figure 2: Circuit for adding 1 K bytes of programmable memory to the SDK-85 computer.

jumpers installed in the appropriate sockets on the board. Figure 1 illustrates the circuit connections for adding a 2708 to the basic board once the address lines have been brought out. Similarly, figure 2 details the circuit for adding 1 K bytes of programmable memory. The suggested placement of these components is in the prototype assembly area provided on the board. Photo 3 demonstrates a viable placement and photo 4 shows how much wiring will be required to make these modifications.

If wired as described, the system memory will be mapped as in table 2.

Computer Software

As stated in the first article in this series,

Hexadecimal Memory Location	Description
2000 thru 20FF	original on board programmable memory
3800 thru 3C00	1 K byte programmable memory expansion
0000 thru 07FF	2 K byte read-only memory monitor
3000 thru 3400	read-only memory expansion area

Table 2: A memory map of the SDK-85 computer after additional memory is added.

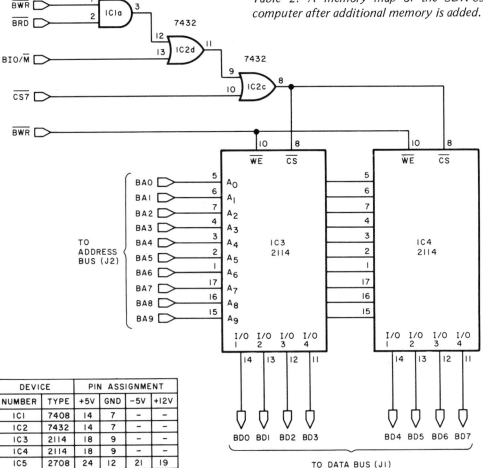

DEVICE		PIN ASSIGNMENT			
NUMBER	TYPE	+5V	GND	−5V	+12V
IC1	7408	14	7	−	−
IC2	7432	14	7	−	−
IC3	2114	18	9	−	−
IC4	2114	18	9	−	−
IC5	2708	24	12	21	19

TO DATA BUS (J1)

Photo 3: Left side of the SDK-85 board contains a prototyping area which, in this case, has been used for the extra memory circuitry. The photo shows an Intel 2716 2 K byte erasable read-only memory which was used because of a desire for single supply power backup. The software only requires 1 K bytes and can use a 2708 instead. Scotchflex connectors to the right of the 2114 and 2716 attach the SDK-85 to the external sensor inputs and control output.

Photo 4: Attachment of additional memory can be easily done using wire wrap techniques. This photo demonstrates the complexity of the connections.

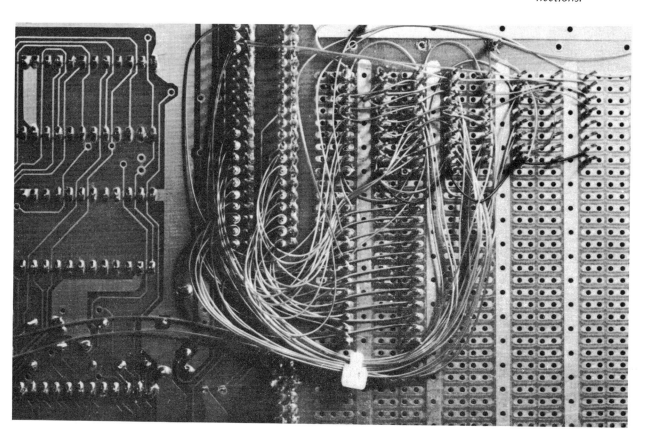

the two major factors involved in developing the software for the home security system are simplicity and flexibility. Simplicity will lead to a straightforward implementation of the design during the coding process and will greatly reduce the time required to debug the code and get on the air.

This approach requires that a considerable amount of time be spent before one line of code is written. During the conceptual phase, an overall system logic is developed. Then the designer begins the task of defining the tables, files, records and logic requirements in detail.

Iterative reflection allows for the development of a "simple" design which is clear and easy for others to follow. In addition, the designer will find that upon completion of

this process, the various software modules will have virtually coded themselves, greatly reducing the time required to code and implement a given design.

System Overview

Figure 3 illustrates the flow of information in this security system.

The cold start procedure causes the system to initialize the sensor state transfer table, delay timer file and time of day file. This procedure requires time of day entry through the keyboard in hours and minutes (24 hour clock) so that the system will be able to activate events in the proper sequence. Upon completion of this procedure the cold start initializes the digital input or

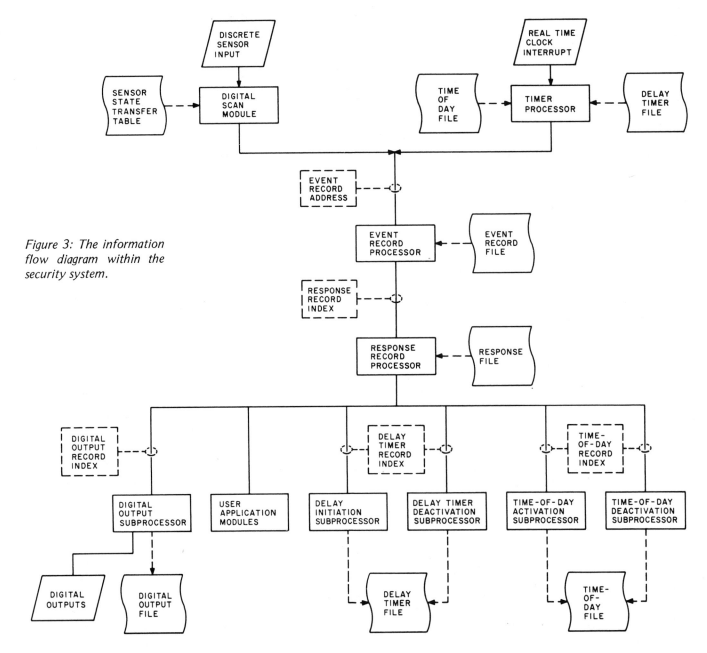

Figure 3: The information flow diagram within the security system.

sensor states, activates the real-time clock interrupt, and starts scanning the sensors. (Intruders beware, for the security system is now active!)

The system continually monitors the state of the various sensors via the digital scan module; and periodically (once a second) checks to see if any time events are to be initiated. In order to demonstrate how a timed event function is processed, let's assume that one of the time of day records initiates the sequence that tells the system to turn the percolator on at 0600. As the system processes each interrupt from the real-time clock, the time in this record will be compared to the current time of day. When they are equal, the events leading to a freshly brewed cup of coffee would be under way.

The first step in this example is effected by the timer processor which extracts an event record address from the time of day record and passes it to the event record processor. The event record processor assumes control, and using the record address passed to it, obtains from the event record file the list of response records to be processed for this event.

Only one response record is required. However, for more complex functions the event record can be structured to execute up to 255 different responses to an event. The response record index, which the event record processor obtains from the event record, is now transferred to the response record processor.

The response record processor, using the index supplied, obtains a record from the response file. In our example this record directs the processor to activate the digital output subprocessor and passes the index of the digital output record to be used in turning on the percolator. Having accomplished this function the subprocessor returns control to the event record processor, which checks to determine if any further responses are required. If so, it will initiate the response. In our example no further response is required and control is returned to the timer processor. This completes processing of those delay and time of day records requiring servicing. This same procedure is used in the servicing of delay timer records whose delay time has expired.

The servicing of a sensor event is initiated by a different mechanism. However, once control is transferred to the event record processor the servicing of the event follows the same sequence as our percolator example. The event record processor will return control to the digital scan module when it completes initiation of all the responses associated with the event.

To illustrate the processing of a sensor detected event, let's assume that you have installed a photodetector at the entrance to your driveway. When this sensor is activated, an exterior light in your home is to be turned on.

The digital scan module is continually scanning the state of the sensors connected to the input ports of the security system. When this module detects that the photo-electric device has been tripped, the sequence that turns on the light is initiated. Using the information it has, the scan module obtains from the sensor state transfer table the record associated with the driveway sensor. This record contains two entries: one for the sensor going to a set (1) state, and one for a reset (0) state. These two entries are provided so that event processing can be initiated on a transfer to a set state, a transfer to a reset state, or both.

Assuming that a reset state triggers an event, the light beam is broken and the signal goes to a logic 0. The digital scan module extracts from the sensor state transfer table the address of the event record to be processed and transfers control to the event record processor. This results in the turning on of the exterior light.

These two examples should illustrate the manner in which the software modules interact with one another. A detailed description of the functions of each module and the structure of the various records, tables and files used in the system follows.

Cold Start

The cold start module requires some special tailoring for each system. It is the responsibility of this module to start the operation of the system.

There are a few basic functions which this module must perform (see figure 4). The primary functions are to transfer the state transfer vector table, time of day file and delay time file from erasable read-only memory to preset location in programmable memory. The records in these files and tables contain volatile fields (data which will be periodically updated).

Another major function of the cold start module is obtaining the current time of day from the person initiating the system. The user enters this data via the keyboard located on the SDK-85 printed circuit board. (The SDK-85 is a good choice for this system because it provides the basic requirements for data entry and display.) The time of day as entered is displayed on the 6 digit LEDs and updated periodically by the real-time clock.

The two remaining functions necessary

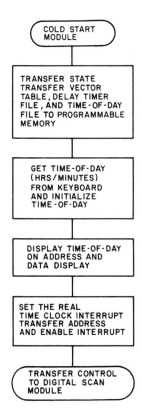

Figure 4: Cold start logic module.

to the cold start process are the initiation of the digital input state table and enabling of the real-time clock interrupt. The detection of a digital event is the result of a comparison of the current state of a digital output with its previous state. Therefore, the initial state must be explicitly set. The state table for all digital or sensor inputs resides in volatile memory and must be set to zero to avoid ambiguity caused by powering up the system.

Since the SDK-85 provides the user with a vectored interrupt capability, the final function of the cold start module is to set the final timer (user vectored interrupt) vector address. This is accomplished by placing a jump instruction at a previously defined address followed by the address of the real time clock interrupt module, ie: JP RTC + JUMP TO REAL-TIME CLOCK.

After initiating the interrupt transfer address, the cold start module enables the real-time clock interrupt and transfers control to the digital scan module.

Digital Scan Module

The responsibility for detecting an intruder, fire, smoke, etc, lies with the digital scan module. This module continually monitors the state of the digital sensor inputs and when it detects that an input has changed state (gone from a set (1) to a reset (0) condition, or a reset (0) to a set (1) condition), initiates the processing of the event.

As can be seen from the flowchart (figure 5), the digital scan module reinitializes itself after each scan. As it scans the current status of the system sensors, the input data is compared bit by bit until all input bits have been processed. If the module determines that no change has taken place in the state of the sensor, the next bit in the sensor state table is processed. This function continues until all sensor states have been checked, at which time the process is started over.

If, however, the above mentioned technique detects that a sensor has changed state, the state history table is updated to reflect the new state of the sensor. The sensor state transfer record is extracted from the sensor state transfer vector table (see table 3).

The sensor state transfer vector table contains two fields for each sensor or digital input. These two entries comprise one record within the table or file, with each digital input requiring one record. The first entry or field within a sensor record contains the event record address to be passed to the event record processor upon a transition to a set (1) condition by the sensor, and the active/inactive flag (bit 7 of the upper address byte). The second field or entry contains the same information as the first, but is used when a transition to a reset (0) condition has taken place. If the active/inactive flag is reset (0) the entry is considered to be active and further processing may take place; if the flag is set (1) no further processing is allowed.

In the event that the sensor has undergone a transition to a set (1) condition, the first entry of the sensor state transfer vector record is obtained from the table, else the second entry is obtained. Then the active flag is examined. If this flag is found to be reset (0), the event record address is transferred to the event record processor for additional processing.

However, if the flag indicates the entry is inactive (flag equals 1), no further processing of the transition event will take place and the digital scan module will continue to process inputs in its normal manner.

As you may have noticed, the manner in which inputs are processed to completion is sequential in nature. That is, when a transition of a sensor requires processing, all processing associated with that event is completed before the next sensor input is processed. This is possible because the processing of a given event will not substantially delay the processing of the next sensor, since no appreciable I/O processing is required. One other feature afforded by this design is that one or more sensors may use the same event record since after the digital input processing there is no further need to maintain the identity of the sensor undergoing the transition.

Since the processing of all inputs is sequential, the user should not incorporate lengthy input or output procedures in a user supplied application module.

Event Record Processor Module

As stated in the digital scan module description, the event record processor is activated when the digital scan module detects a state transition which requires processing. This module is also activated when an active delay timer (see timer interrupt processor module) or time of day event occurs.

Table 3: Sensor state transfer vector table format.

	Byte	Description
Record 1 {	0-1	Event Record Address for RESET to SET State Transfer for Digital Input 1 (High Order Bit=Active Flag)
	2-3	Event Record Address for SET to RESET State Transfer for Digital Input 1 (High Order Bit=Active Flag)
Record n {	3-4	Event Record Address for RESET to SET Transfer for Digital Input 2 (High Order Bit=Active Flag)
	5-6	Event Record Address for SET to RESET Transfer for Digital Input 2 (High Order Bit=Active Flag)
	*	
	*	
	*	
Record n {	(2n−1)−2n	Event Record Address for RESET to SET State Transfer for Digital Input n (High Order Bit=Active Flag)
	(2n+1)−(2n+2)	Event Record Address for SET to RESET State Transfer for Digital Input n (High Order Bit=Active Flag)

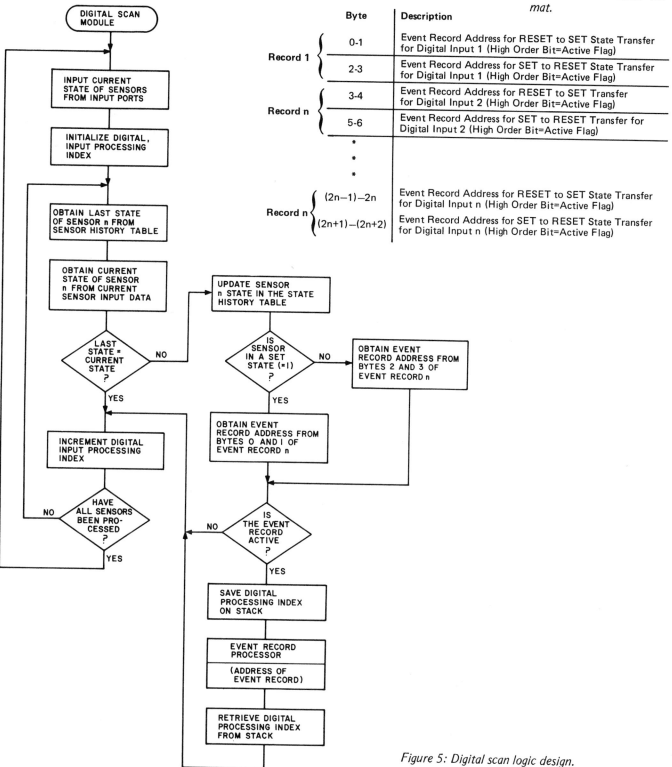

Figure 5: Digital scan logic design.

33

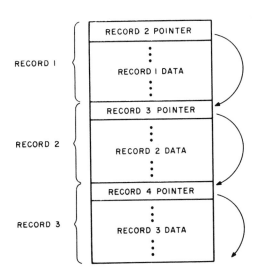

Figure 6: Representation of chained records in memory space.

This structure of the event record dictates the manner in which the record being processed must be accessed. Generally all records within a file are of fixed length and can be referenced by an index which serves as a relative pointer into the file to access the record in question. However, when the records within a file are of variable length, an absolute or relative address must be provided to obtain access to the record in question. This address may be obtained via two different techniques. The first technique requires that all records in a file be "chained" together. This means that there is a data field in the first record which points to the second record in the file, which in turn points to the third record, etc (see figure 6).

To access a record in this type of file the record index is supplied to the access module which then sequences the chain down (or up) until the proper record is located. This structure is used when a record search is required to extract data from one or more records in the file, but in the case of this system offers no advantages.

The second technique, and the one employed in this system, requires a directory of record addresses (see table 3), so that the module detecting the requirement for event

The address of the event record requiring servicing is given to the event record processor. The event record (see table 4) contains the addresses of the response records to be processed when a digital scan (sensor), delay timer or time of day event occurs. The records in the table or file of event records are of varying length in order to allow one or more specific responses to be associated with any event.

Figure 7: Event records are accessed through sensor state transfer vectors.

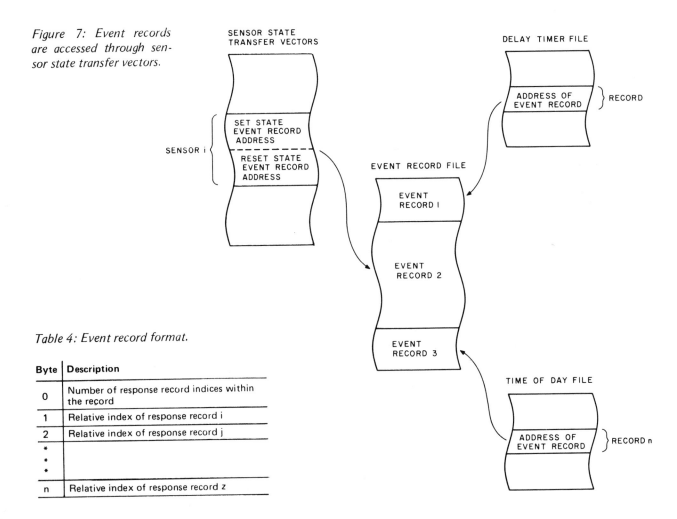

Table 4: Event record format.

Byte	Description
0	Number of response record indices within the record
1	Relative index of response record i
2	Relative index of response record j
*	
*	
*	
n	Relative index of response record z

processing will have access to the event record associated with the event (see tables 7 and 8) as is done with the timer delay and time of day records. An event record can therefore be isolated via one of three sources in the security system: the sensor state transfer vector table, the time of day records and the delay records (see figure 7).

Since an event record may contain more than one response directive, the first data field (byte 0) of the record contains the number of response records (see table 5) associated with the event in question. Those fields following within the record contain the relative indices of the response records that require processing.

Upon entry, the module saves the event record address in a temporary working area and then extracts from the event record the number of response records to be processed (see figure 8). This data will then be used to initialize the event record processor counter and the initial response record address will be calculated:

Response Record Address =
(Response Record Index − 1) × 3
+ Base address of Response Record File

This response record address is then passed to the response processor for execution. When completed, control is returned to the event record processor. If additional records require processing, the next effective response record address is calculated and control is transferred to the response record processor. If all records have been processed, control is returned to the module initiating the request for service, either the digital scan module or the timer interrupt module.

Response Record Processor

The function of the response record processor (figure 9) is to determine from the response record the type of action required, and to activate the appropriate response subprocessor. The activated subprocessor will effect the final response and then return control to the event record processor.

There are six basic responses associated with an event. These responses are defined by the response record (table 5) and are:

- Activation of a digital output
- Execution of an application module
- Initiation of a delay timer
- Deactivation of a delay timer
- Activation of a time of day record
- Deactivation of a time of day record

The user should be aware of the function of each subprocessor and certain idiosyncrasies. The processing of a digital output

is the function of the output subprocessor. This processor will, upon activation, receive from the response processor the index of the digital output record (table 6) it is to process. The digital output record contains the information necessary to effect the desired output. The current output state of each output bit of the port is maintained in the port's state word in the output port state file. Each set (1) bit in the state word represents an activated output, and each reset bit represents a deactivated output.

The output port identifier (byte 0) of the digital output record is used to extract the state data from the output port state file.

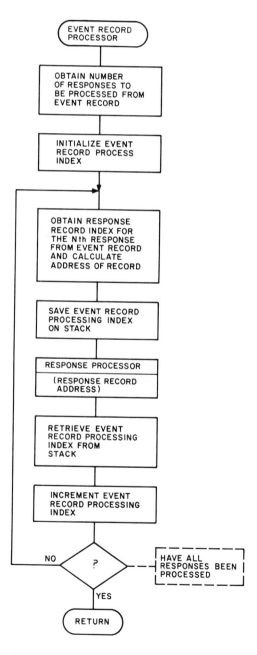

Figure 8: Event record processor logic diagram.

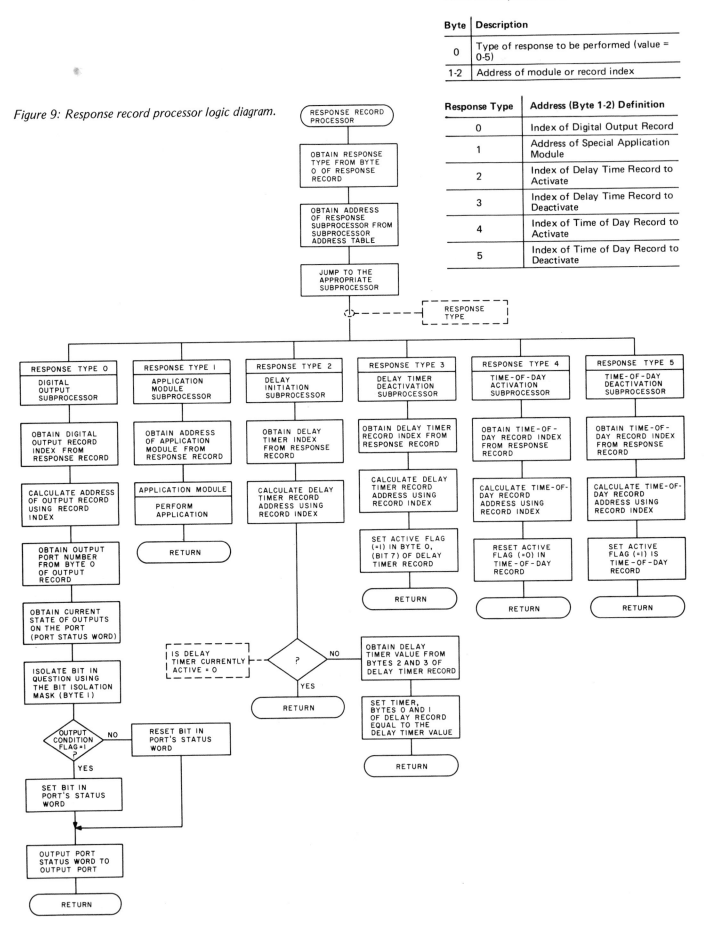

Figure 9: Response record processor logic diagram.

Table 5: Response record format.

Byte	Description
0	Type of response to be performed (value = 0-5)
1-2	Address of module or record index

Response Type	Address (Byte 1-2) Definition
0	Index of Digital Output Record
1	Address of Special Application Module
2	Index of Delay Time Record to Activate
3	Index of Delay Time Record to Deactivate
4	Index of Time of Day Record to Activate
5	Index of Time of Day Record to Deactivate

Byte	Description
0	Output Port for Output
1	Bit Isolation Mask
2	Output Condition Flag

Table 6: Digital output record format. If the output condition flag is 0, the output bit is turned off (0). If the output condition flag is 1, the output bit is turned on (1).

The bit to be operated upon is then isolated using the bit isolation mask (byte 1), and the state of the bit is set or reset as directed by the output condition flag (byte 3). The output bit is then merged back into the port's state word, stored in the output port state file, and the port's new state is output. This procedure allows the output subprocessor to effect a state change on a device without affecting the state of any other device connected to the port in question.

User supplied application modules may be activated by the response record processor by setting the response type equal to 1 in the response record and providing the initial execution address of the module in bytes 1 and 2 (low order in byte 1) of the response record. This will cause the response record processor to execute a jump to the specified execution address. Since all response subprocessors must return control to the event record processor, the user must exit his module with a return instruction.

The initiation of delay timers (used to delay a response) is the responsibility of the delay initiation subprocessor. This module is, upon activation, given the index of the delay record (table 7) to be initiated. Upon initiation the subprocessor determines if the delay timer record is currently active. In the event that the time is active, control is returned to the event record processor, leaving the delay timer in its current state. In other words, an active delay timer is not reactivated, nor is its delay time reset to the initial value indicated in the record. However, should the subprocessor determine

Table 7: Delay time record format. If the active flag is 0, the timer is active. If the active flag is 1, the timer is inactive.

Byte	Description
0-1	Active Flag (Bit 7, Byte 0) Delay Time (Seconds)—(Bit 6-0, Byte 0, Bit 7-0, Byte 1)
2-3	Timer Activation Value (Maximum value = 32,767 seconds)
4-5	Address of Event Record associated with Delay Record

Byte	Description
0	Active Flag (Bit 7) Hours (Bit 6-0) (0-23 hours)
1	Minutes (0-59 minutes)
2-3	Address of Event Record Associated with Time of Day Record

Table 8: Time of day record format. If the active flag is 0, the record is active. If the active flag is 1, the record is inactive.

that the record is inactive, the active flag is reset, the timer's activation time is transferred to bytes 0 and 1 of the record, and control returns to the event record processor.

The deactivation or disabling of active delay timers is the responsibility of the delay timer deactivation subprocessor. This module is passed the index of the delay timer record to be deactivated. Using this data the address of the record in question is determined and the active flag (bit 7, byte 0) is set equal to 1, thereby disabling the timer. This same procedure is also used to deactivate a time of day record (table 8).

The activation of a time of day record is performed by the time of day activation subprocessor. This module is, upon activation, given the index of the time of day record (table 8) to be initiated. Upon initiation the subprocessor will determine the address of the record and the active flag (bit 7, byte 0) will be reset (0), thereby enabling the time of day record. The subprocessor will then return control to the event record processor.

Timer Processor

As stated in part 1 (page 6), a real-time clock with a frequency of 1 Hz was used to provide the time base for the system. This periodic signal causes the processor to generate an interrupt, thereby causing the current contents of the program counter register to be saved on the stack and control to be given to the timer processor (figure 8). The functions of the timer processor are:

- Handle real-time clock interrupts
- Display current time of day
- Process delay timer records
- Process time of day records

Since the timer process is interrupt driven and capable of assuming control during any function of the system, it must be assured that the program it interrupted may resume execution. To assure this capability, the timer processor immediately upon activation saves the registers of the interrupted program on the stack. Using this technique it is possible upon completion of the time processor to restore the registers to the state they were before the interrupt and to return control to the interrupted module.

Upon completion of the saving of the interrupted modules registers, the time of day being maintained by the computer is updated and displayed. Hours are displayed in the upper two digits of the address display, minutes in the lower two digits of the address display, and seconds in the data display. To avoid any ambiguity in discerning AM from PM, time is maintained using a 24-hour clock. In this system, then, 0 hours, 1 minute, 0 seconds corresponds to 12:01 AM; 13 hours, 0 minutes, 0 seconds corresponds to 1 PM.

Figure 10: Timer processor logic diagram.

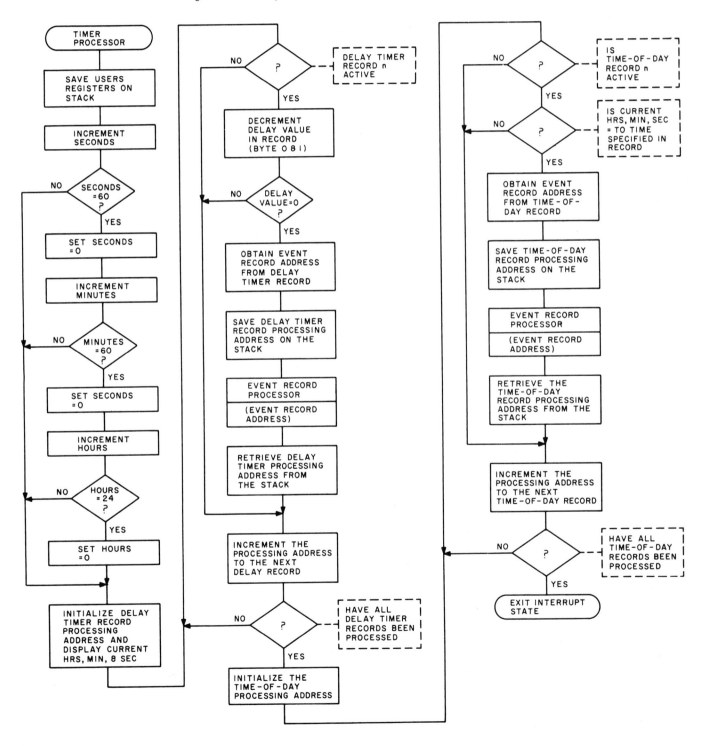

The timer processor scans the file of delay timer records (table 7) to process those records which are active. The function of the delay timer record is to provide a delay of a preset number of seconds before processing an event record. To demonstrate how this function works, let's assume that one wishes to activate an audible alarm 50 seconds after a particular sensor has been tripped. You would structure the event and response records so that the tripping of the sensor would activate the delay timer record you were associating with this event. The activation of the delay timer causes the active flag to be reset and the timer activation value (bytes 2 and 3) to be transferred to the delay time. In our example, this will cause the value 50 to be loaded into bytes 0 and 1 of our delay timer value.

As each real-time clock interrupt causes the timer processor to scan the file of delay timer records, it will find that our record is active. Once the processor determines that a record is active it will decrement the current delay time remaining and then check to see if the time remaining is zero. If there is still delay time remaining (time greater than zero) no further action is taken.

However, if there is no time remaining (time equals zero) the active flag is set, the current registers saved, the address of the event record is extracted from the delay timer record, and control transferred to the event processor. The sequence of operations performed from this point will directly result in the audible alarm being turned on.

The processing of the records in the time of day file (see table 6 for record format) is performed upon the completion of all delay time records. Time of day records are processed in a manner very similar to that described for delay timer records. When an active record is encountered during the scan of the time of day file, the current time of day being maintained by the system will be compared to the time of day specified in bytes 0 and 1 of the record. Should these times be identical, the records flag is set to 1 and the processing of the event record specified in bytes 2 and 3 is initiated as described above. After servicing all records in the time of day file, control is returned to the interrupted modules.■

Build a Computer Controlled Security System for Your Home: Part 3

There are many security systems on the market. From the simple $10 door buzzers which signal forced entry to elaborate professionally installed Rollins and ADT systems, their purpose is singular: give the occupant advance warning of an emergency condition. Protection of property is a secondary benefit of the more sophisticated alarms. Ultimately it is the overall complexity of the security system that defines how much coverage is attained in each of these areas.

Forced entry, prowler, and fire detection are but three possible events for which people buy alarms. A $15 smoke alarm alerts the occupants, who rush out of the house in the nick of time but stand watching the house burn because they didn't have time to call the fire department. Similarly, a prowler breaks into a home when the family is out. He has grown accustomed to the regular pattern of timer controlled lights after observing the house for a few nights and immediately disables the alarm horn upon entry. Had the occupants been home they of course would have been alerted to the break-in, but that was not the thief's intention.

To provide full protection, the ultimate security system should discourage intrusion, monitor all potential emergency situations and have the intelligence to initiate a preset series of actions should the alarming event ever occur. Combining all these elements

into a single computer controlled security system is the subject of this and the two preceding articles (pages 16 and 26). With the control system proposed and developed in this series, the user will have more of a process control computer than a burglar alarm. Except for the detection of the alarming event itself, a system designed to discourage intrusion and automatically respond to certain situations cannot be configured solely as a passive detection unit. The programs which this computer executes either in response to timed events or sensor inputs amount to a process control situation.

The computer controlled security system outlined in these three articles has the capability (presuming you have wired the same output controls as I have) to detect cars or people approaching a residence and track them by sequentially turning on flood lights aimed to cover the perimeter of the house, or give an audible warning in daylight situations. It can also control the power to AC appliances (TV, lights, stereo, etc) either to simulate occupancy or provide the luxury of remote control. This same software allows preset responses to water and temperature sensors for the control of wood heating, water circulating stoves or simply to turn on a pump in the basement. Should any perpetrator be ignorant enough to break in even after sufficient warning, the system has all the usual bells and whistles, an automatic telephone dialer, a hidden video tape recorder to obtain evidence, and, finally, a separate communications link to your neighbor.

The first article presented a general outline of the control system, the types of inputs available, and proposed responses. The second article described the hardware modifications to the SDK-85 computer used as the controller and provided an extensive description of the control and data acquisition software. With these three articles, more advanced readers should have enough information to configure a similar control computer using any microprocessor.

The final installment in the series presents examples of discrete input sensors to monitor temperature, moisture, motion, fire, and smoke. To allow the computer to make the proper response to such inputs, designs for flashing lights, strobes, siren, and AC remote control interfaces are also detailed. Finally, just to alleviate any lingering apprehensions over what appears to be a complex software algorithm, we will trace the flow of an alarm input as it is processed by the security system and demonstrate the versatility of table-driven software in this application.

The Alarm System Is Only as Good as Its Input Sensors

The SDK-85 has 38 bits of parallel I/O (input/output). In this application 18 of these bits have been set aside to control outputs, while the remaining 20 are used solely for input. Because the computer is composed of TTL (transistor-transistor logic) circuitry, any input to it must also be TTL compatible. It is a further requirement that all signals be discrete in nature. That is, they change state from a 1 to a 0 or a 0 to a 1 level when the set point is attained. Analog inputs are not beyond the capabilities of the software, but no analog to digital interface has been built into the system.

Analog Input Sensors

Temperature and voltage are important analog parameters which any sophisticated alarm system requires if it is to monitor freezer or furnace room temperature and brownout conditions. The circuit of figure 1 can be used to read temperature setpoints. IC1 is a special integrated circuit having an output voltage proportional to temperature.

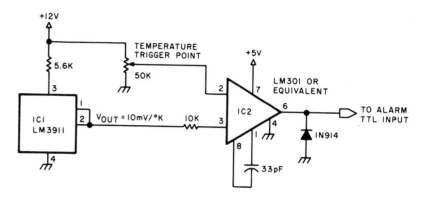

Figure 1: Temperature setpoint indicator circuit. IC1 is a special integrated circuit whose output voltage is proportional to temperature. The circuit feeds into an Intel SDK-85 single board computer, which accepts only TTL (transistor-transistor logic) level inputs. Op am IC2 is used as a comparator to convert the output accordingly.

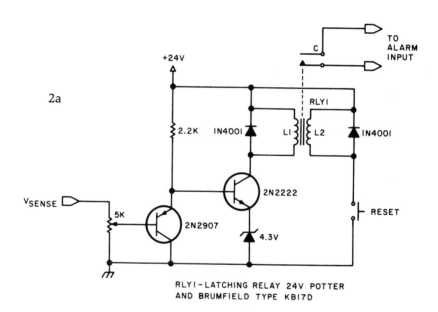

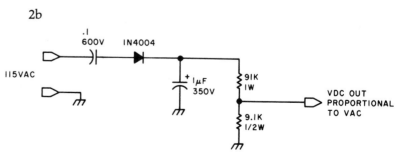

Figure 2: Circuit which can be used to monitor any DC voltage between 0 and 24 V. A latching relay is used to latch on when a voltage transient occurs.

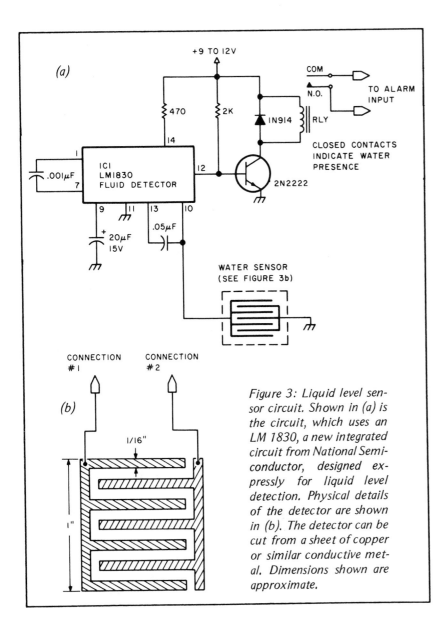

(a)

ICI
LM1830
FLUID DETECTOR

+9 TO 12V

COM
N.O.
TO ALARM
INPUT

RLY

1N914

2N2222

CLOSED CONTACTS
INDICATE WATER
PRESENCE

470   2K

.001µF

+ 20µF
15V

.05µF

WATER SENSOR
(SEE FIGURE 3b)

(b)

CONNECTION
#1

CONNECTION
#2

1/16"

1"

Figure 3: Liquid level sensor circuit. Shown in (a) is the circuit, which uses an LM 1830, a new integrated circuit from National Semiconductor, designed expressly for liquid level detection. Physical details of the detector are shown in (b). The detector can be cut from a sheet of copper or similar conductive metal. Dimensions shown are approximate.

Since the SDK-85 can respond only to TTL inputs, op amp IC2, configured as a comparator, changes logic levels when the output of IC1 equals the setting of the temperature trigger pot. IC2 is powered by a 5 V power supply so that the output is TTL compatible. If a multiple array of temperature transducers is to be constructed, use an LM339 quad comparator instead of four LM301s to reduce wiring complexity. Another analog circuit, shown in figure 2, can be used to monitor any DC voltage between 0 and 24 V. It could be used to monitor the battery backup supply to the computer or some other important parameter. Since voltage fluctuations are often significant and desirable to catch, the circuit uses a latching relay to keep it in the set condition once triggered. If that is not a desirable feature, then replace RLY1 (relay 1) with a nonlatching type. A relay should continue to be used, however, because it isolates the computer from the monitored voltage sources.

Discrete Input Sensors

Actually, all inputs to the computer are discrete, and this need not be a separate category. The outputs from the sensors are primarily contact closures for a reason I'll explain later. Some of the important ones worth considering are liquid level or moisture, smoke and fire, and ultrasonic and infrared interrupted beam motion detectors.

A liquid level sensor suitable for detecting basement flooding so that a sump pump can be turned on is illustrated in figures 3a and 3b. This circuit uses a new integrated circuit from National Semiconductor specifically

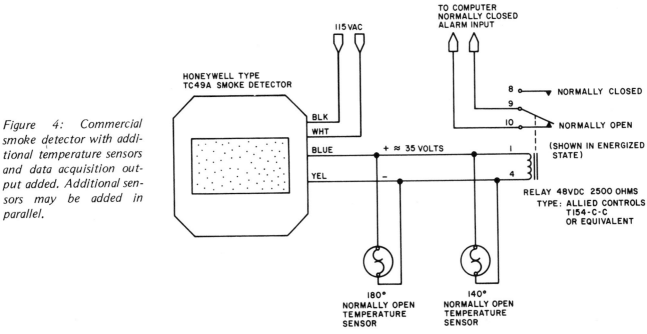

Figure 4: Commercial smoke detector with additional temperature sensors and data acquisition output added. Additional sensors may be added in parallel.

115 VAC

TO COMPUTER
NORMALLY CLOSED
ALARM INPUT

HONEYWELL TYPE
TC49A SMOKE DETECTOR

BLK
WHT
BLUE
YEL

+ ≈ 35 VOLTS
−

8   NORMALLY CLOSED
9
10   NORMALLY OPEN

(SHOWN IN ENERGIZED
STATE)

1

4

RELAY 48VDC 2500 OHMS
TYPE: ALLIED CONTROLS
T154-C-C
OR EQUIVALENT

180°
NORMALLY OPEN
TEMPERATURE
SENSOR

140°
NORMALLY OPEN
TEMPERATURE
SENSOR

designed for this purpose. As configured, the relay contacts will close in the presence of water or condensing moisture. The circuit is extremely sensitive and can be used with practically any conductive liquid.

Probably the most important sensors on the security system are the ones that detect fire and smoke. A commercially available ionization type smoke detector, shown in figure 4, has been configured to interface to the computer. The AC powered TC49A by Honeywell is ideally suited for this purpose, since it is designed for parallel attachment to other sensors. In this instance any number of normally open temperature sensors can be attached between the blue and yellow output leads of the device. Should any of these sensors or the smoke detector itself detect an alarm condition, the 35 V normally present on these wires will drop to 2 V. The relay which had previously been in an energized state will open, signaling the event to the computer. Protection must be provided during power failures, however, so that the computer (which has an emergency supply) does not detect a false alarm. If the program which scans the smoke alarm also checks the power failure sensor, positive results should be obtained. In any case it is always a good idea to have a battery powered smoke detector also within the residence.

Note that the majority of sensor designs presented in this article can be used independently. The device normally activated when the sensor signals an alarm condition can be attached and directly controlled through another parallel set of relay contacts. An example would be the water detector that automatically turns on the sump pump. Requiring the computer to receive the signal, process the control record, and turn on an AC powered pump would be a waste of wire. The computer need know only that the high water mark has been reached to notify the residents on the display panel: it does not have to control it as well. Before you string a mile of wire through the house, consider what functions really need "computer" rather than "local" control.

The two remaining special input sensors are related in purpose. Both are used to detect an object or person passing between two points, and both use interrupted beam sensing techniques. One is an infrared light beam and the other is ultrasonic. The light beam circuit is shown in figure 5 and the ultrasonic circuits are illustrated in figure 6. The range of the infrared unit is about 10 feet without a lens and as much as 50 feet with proper ambient light shielding and a focusing lens. No focusing was tried on the ultrasonic unit, but 25 feet was easily achieved.

Testing and alignment of the ultrasonic transmitter can be tricky, while the infrared is simply a mechanical alignment consideration. First, the transmitter must be tuned to resonance. The nominal frequency of the ultrasonic transducers can be 34 to 42 kHz; they should be bought in pairs. An oscillo-

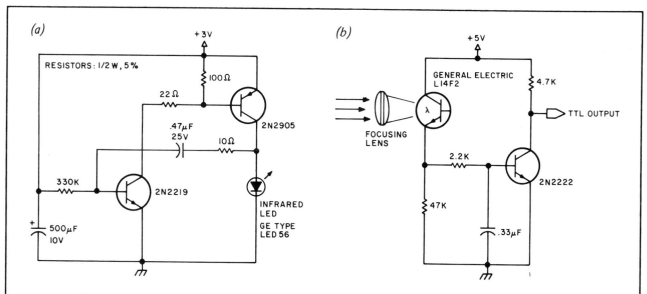

Figure 5: Intrusion alarm infrared transmitter and detector. Shown in (a) is a low voltage pulsed LED (light emitting diode) transmitter; the light beam receiver with optional focusing lens is shown in (b). The lens extends the range of the unit from approximately 10 to 50 feet. Anyone or anything breaking the beam will alert the security system.

scope should be put across the transducer in the transmitter circuit when power is applied. Coils T1 and T2 should then be adjusted to produce the greatest amplitude across the transducer. The usual value is about 30 V peak-to-peak and the frequency should be the nominal F_0 listed for the part.

Once the transmitter is tuned, place it about 2 feet in front of the receiver. Adjusting the center frequency adjustment pot should cause the relay to pull in and the LED (light emitting diode) to light. The transmitter and receiver can now be placed across a driveway or large room.

Wireless Inputs

So far we have only discussed input sensors which are directly wired to the connectors of the SDK-85 computer. If a reed switch were attached to the garage door a wire must be run from it to the computer. In a larger house this can amount to a lot of wire and can extend system construction time. One possible solution is to use a wireless transmitter and receiver between the computer and remote points within the house.

Homebrew wireless transmitters, while cheap, suffer from a lack of reliability. They are not being considered for this application because there is a commercial unit available which is both cost effective and reliable. The particular device is the Norelco Home Patrol wireless burglar alarm system available in most discount stores for about $200. It consists of a receiver, four contact-closure-activated transmitters, and a smoke

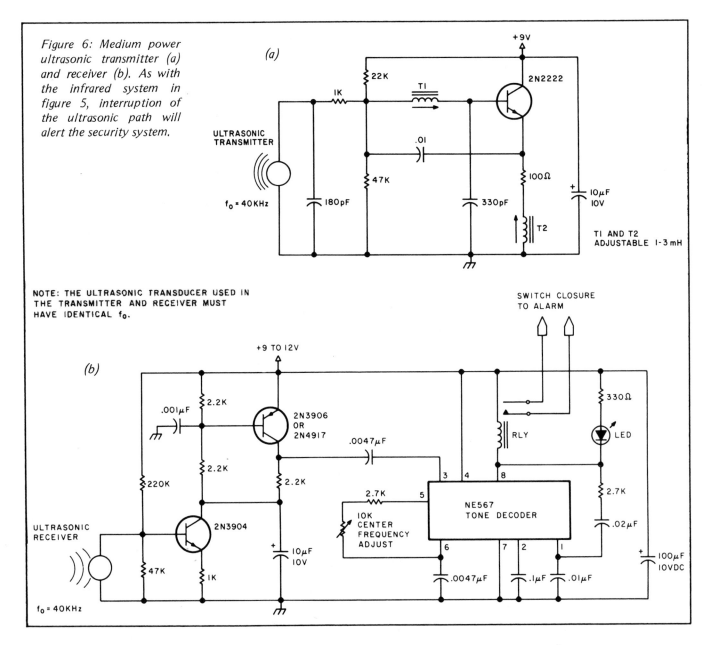

Figure 6: Medium power ultrasonic transmitter (a) and receiver (b). As with the infrared system in figure 5, interruption of the ultrasonic path will alert the security system.

detector with built-in transmitter. Photo 1 shows the components of the system and figure 7 details how each of these separate transmitters can be expanded to cover a wider area.

The Norelco receiver has four separate output channels designated as fire, intrusion, car, and miscellaneous. Transmitters are supplied for intrusion and fire only — transmitters for the other two channels must be purchased separately. Photo 2 shows one of these devices.

Photo 3 shows the output connections of the receiver and figure 8 illustrates the type of interface which must be constructed to convert the 0 and —15 V Norelco receiver outputs to be TTL compatible. Only then can the security system be aware of these remote alarm inputs.

Security System Outputs

Once an alarm condition has been detected or the event processor activated, the security system responds accordingly. Whatever the cause, the output will be a TTL change of state which can be used to drive a mechanical or solid-state relay. Typical output interfaces are warble alarms, high intensity flashers, and strobe lights. They are shown in figures 9, 10 and 11, respectively.

AC output control can be handled in either of two ways: solid-state or mechanical relays. While solid-state relays are definitely the more modern approach, it is very difficult to find control panels incorporating them which are understandable to electricians or which meet local electrical codes. Rather than fight the system it was easier to install a readily available relay control panel as shown in photo 4. The two cabinets on the right are relay cabinets, and the one on the left is the regular breaker box. Each relay enclosure contains six relays. The left relay enclosure controls the six outside light circuits and the righthand enclosure remotely controls six wall outlets around the house. The relays are DC input and can either be controlled directly from the computer or manually from scattered points around the house. Each of the 12 relay outputs requires a separate cable to the outlet or light to be controlled. This is not an inexpensive control method, but it does meet the code and is a convenience once installed.

Two other details left to be considered are the emergency power supply to the com-

Photo 1: Norelco wireless alarm system. The three components from left to right are: wireless smoke detector, wireless door sensor, and master receiver.

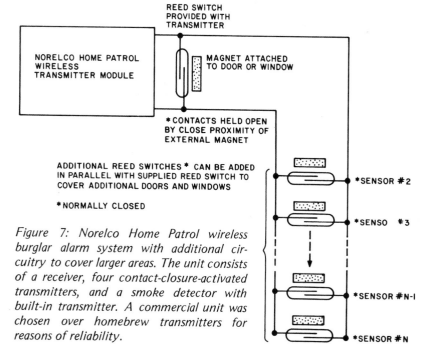

Figure 7: Norelco Home Patrol wireless burglar alarm system with additional circuitry to cover larger areas. The unit consists of a receiver, four contact-closure-activated transmitters, and a smoke detector with built-in transmitter. A commercial unit was chosen over homebrew transmitters for reasons of reliability.

Photo 2: Single wireless transmitter with one read switch attached to it. Multiple reed switches can be attached in parallel to cover a wider area, as described in the text.

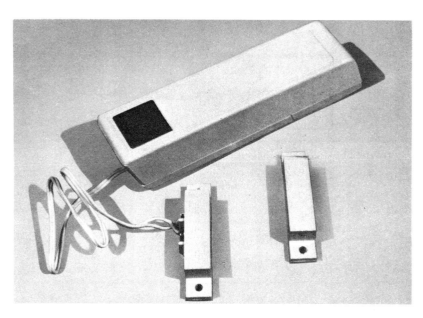

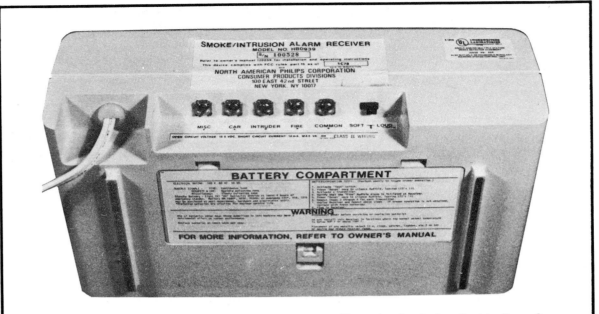

Photo 3: Output connections of the Norelco receiver. Using the circuit described in figure 8, the outputs of the Norelco receiver can be easily interfaced to the computer. There are four distinct output channels of the Norelco unit. As purchased in its basic form it comes with transmitters for fire and intrusion only.

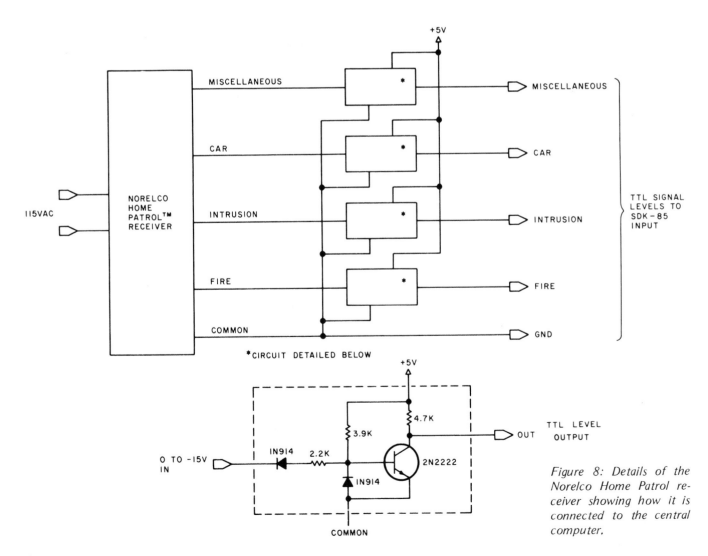

Figure 8: Details of the Norelco Home Patrol receiver showing how it is connected to the central computer.

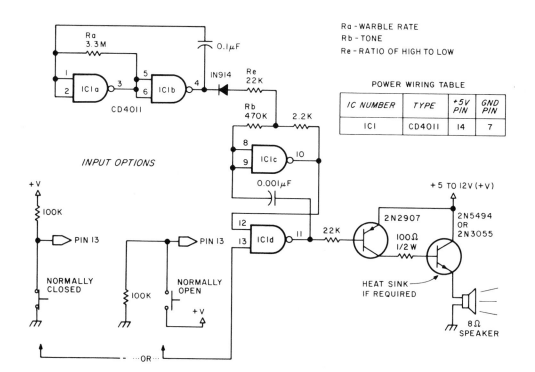

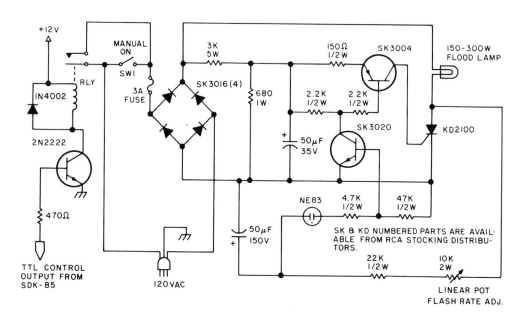

Figure 9: CMOS warble alarm.

Figure 10: 115 VAC incandescent lamp flasher.

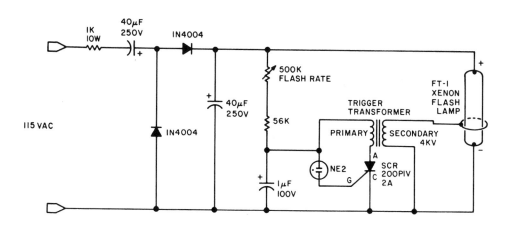

Figure 11: 115 VAC Xenon strobe light.

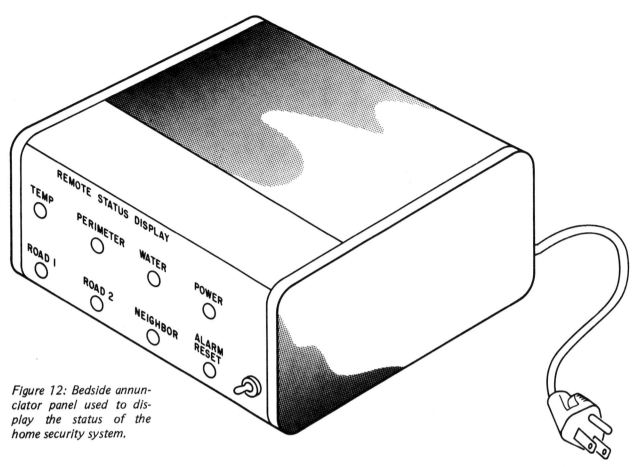

Figure 12: Bedside annunciator panel used to display the status of the home security system.

puter and the display panel located in the bedroom.

The Alarm Status Display Panel

The purpose of the alarm status display panel is to provide the resident with a graphic representation of the status of the security system. Figure 12 is a sketch of a simple annunciator (the picture was omitted because the control panel I finally built bears no resemblance to the purpose we are discussing, and might be confusing since it contains numerous additions to the basic concept).

The circuit (see figure 13) is simply a BCD (binary coded decimal) to decimal decoder driver which is multiplexed by a 3 bit output from the SDK-85. The computer sequentially sends out the codes for the particular lights to be lit. This is done repeatedly and with sufficient speed so that they appear constant. If the annunciator panel is more than 50 feet from the computer, the user may want to consider the addition of the line drivers also described.

Emergency Power Supply

Should power ever be lost in the residence, it is important to maintain the security system in an active mode. To do this a 12 V automobile battery is used to

power the computer *all the time*. The SDK-85 5 V power is derived from the 12 V through a regulator. The EROM (erasable read-only memory) can be either a single +5 V unit such as a 2716; or, if using a 2708, the −12 V can be derived using one of the circuits outlined in my article "No Power for Your Interfaces? Build a 5 W DC to DC Converter" (October 1978 BYTE, page 22). The requirements of such a power supply are maintenance of the 12 V on the battery, recharging the battery as it needs it, and also providing standby power to critical sensors and alarms.

Tracing an Activated Alarm Condition: An Example

Part 2 of this series (page 16) emphasized the software of our computerized home security system. To adequately complete the description of this design, it is necessary to include an example which illustrates the use of table driven software. Included is a listing of the digital scan module and various response modules which should aid in understanding.

Suppose one of the functions of the system is to respond to a smoke and fire detector. First, assume that the sensor in question is wired into the least-significant bit (bit 0)

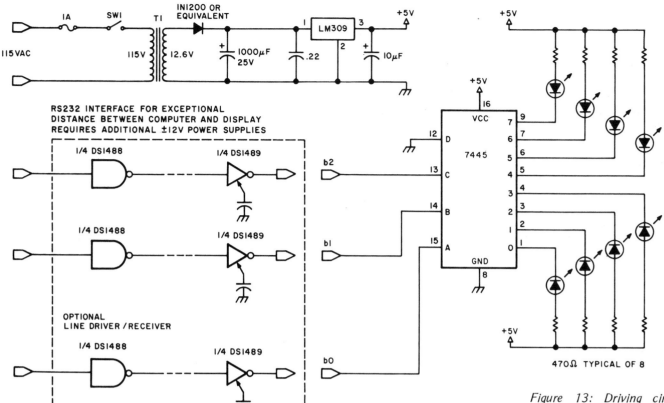

Figure 13: Driving circuitry for the annunciator panel (see figure 12). A BCD (binary coded decimal) code from the computer is converted to decimal by IC7445 to drive the LEDs on the panel, which are driven sequentially with sufficient speed so they appear to be lit constantly. An optional RS-232 interface for driving the circuit at some distance from the computer plus a 5 V power supply are also shown.

on port 0. Response to this sensor is on a reset-to-set (0 to 1) transition, and that on a set-to-reset (1 to 0) transition no action is to be taken. As detailed in part 2 of this series, the initial state change is detected by DIGSCN (the digital scan module) which uses the information contained in XFVE (the sensor state transfer table) to initiate the processing of a digital event. If you recall, the sensor state transfer table is comprised of four byte records, with one record being required for each digital input:

Sensor State Transfer Record

Bytes 0 and 1: event record address for reset-to-set transition plus active/inactive flag (bit 7);
Bytes 2 and 3: event record address for set-to-reset transition plus active/inactive flag (bit 7).

The smoke and fire detector connected to bit 0 input uses the first record in the sensor state transfer table to initiate the required responses. Since no action is required for a set-to-reset transition, bytes 0 and 1 are set equal to hexadecimal FF. This makes such a transition inactive (active/inactive flag = 1). We do, however, want to process a smoke and fire alarm. Therefore, bytes 2 and 3

must contain the address of the event record associated with the alarm. For this example let us assume that the address of the event record used to process this alarm is at hexadecimal 4214 (therefore byte 2 of the transfer record will equal hexadecimal 14 with byte 3 being equal to 42):

Sensor State Transfer Record for Smoke and Fire Alarm

Byte 0 = FF;
Byte 1 = FF;
Byte 2 = 14;
Byte 3 = 42.

To summarize, the digital scan module detects the transition of the smoke and fire sensor from a reset to a set state. Using the information contained in the sensor state transfer table, the system processes the event by first extracting the address of the event record associated with this alarm and then activating EVPRO (the event processor module).

EVRREC (the event record), which is used by the event record processor, contains the indices to the various response records associated with this alarm. As you may recall, one of the features of this system is its ability to associate several responses to

Photo 4: Electrician Vince Sadosky installing "Touch Plate" control system. Each enclosure (of the two on the right) contains six 15 A latching relays. The relay is a pulse-on/pulse-off type in which the control pulse can come from either a manual pushbutton panel in the kitchen or from the computer.

a single event. The event record is the mechanism which accomplishes this as follows:

Event Record Format

Byte 0: number of responses associated with record;
Byte 1: relative index of response record i;
Byte 2: relative index of response record j;

•
•
•

Initially the response might be to activate the autodialer notifying the police and fire departments of the predicament. But before doing that, let us take a few seconds to check for a false alarm and, if possible, verify the situation. This can be accomplished with the following responses to the alarm. First we do want to initiate the autodialer. So, our first response will be the initiation of a 60 second delay of the autodialer. Next, since we may be in the bedroom and unable to hear the alarm itself, we will initiate an audible alarm — and, so that we can tell the location of the alarm, we will display the location of the alarm on the annunciator panel.

These functions will require an event record containing the indices of the following responses:

- Delay timer activation for the 60 second delay.
- Application task response for driving the annunciator.
- Digital output response for activating the audible alarm.

In this example it is assumed that the response indices are 0, 1, and 2, respectively. The event record for this event will therefore be four bytes in length as follows:

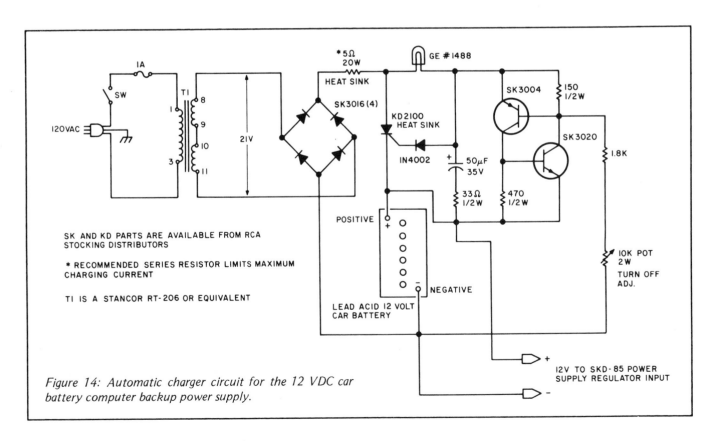

Figure 14: Automatic charger circuit for the 12 VDC car battery computer backup power supply.

Event Record for Smoke and Fire Alarm

Hexadecimal Address

4214 = 3 }	number of responses;
4215 = 0 }	
4216 = 1 }	response record indices for
4217 = 2 }	event.

The event processor will process the entries in the event record sequentially, extracting the response record index and activating RESPRO (the response record processor). Using this index the response processor will obtain RESREC (the indicated record from the response file) and direct the activation of the appropriate subprocessors, which contain the following information:

Byte 0 = response type
{
0 = digital output;
1 = application module;
2 = delay activation;
3 = delay deactivation;
4 = time-of-day activation;
5 = time-of-day deactivation.
}

Bytes 1 and 2 = module address of record index.

For our example the records will be as follows:

Response File

Record 0	{ Byte 0 = 2 Byte 1 = 0 } Byte 2 = 1 }	second delay timer record;
Record 1	{ Byte 3 = 1 Byte 4 = 0F } Byte 5 = 2C }	address of annunciator application module;
Record 2	{ Byte 6 = 0 Byte 7 = 0 } Byte 8 = 0 }	first digital output record.

The first record processed causes DELINT (the delay module) to activate the second record in TIMREC (the delay timer file). These records contain the delay time (in seconds) and address of the event record to be activated when the directed delay has timed out:

Delay Timer Record

Bytes 0 and 1: active flag plus remaining time;
Bytes 2 and 3: number of seconds to delay;
Bytes 4 and 5: address of event record.

In our example this record (prior to activation) will appear as follows:

Bytes 0 and 1: 80, 0 = record inactive;
Bytes 2 and 3: 0, 3C = 60 second delay;
Bytes 4 and 5: 18, 42 = event record at hexadecimal 4218.

After the delay has been initiated, the event processor will notify the response processor to initiate the second response.

Listing 1: Selected 8080/8085 subroutines for the home security system.

```
ZERO      EQU    0
ONE       EQU    1
P1        EQU    0          *PORT 1
P2        EQU    1          *PORT 2
P3        EQU    2          *PORT 3
*
* THE FOLLOWING AREAS ARE TO BE LOCATED IN RAM
*
TEMP1     DB     0
TEMP2     DB     0
TEMP3     DB     0
PORT1     DB     0          * CURRENT INPUT DATA FOR PORT 1
PORT2     DB     0          * CURRENT INPUT DATA FOR PORT 2
PORT3     DB     0          * CURRENT INPUT DATA FOR PORT 3
PORT1H    DB     0          * PREVIOUS INPUT STATE FOR PORT 1
PORT2H    DB     0          * PREVIOUS INPUT STATE FOR PORT 2
PORT3H    DB     0          * PREVIOUS INPUT STATE FOR PORT 3
OPORT     DS     3          * OUTPUT PORT STATUS
DIGIND    DB     0          *DIGITAL INPUT BIT PROCESSING INDEX
XFVE      DS     500        *SENSOR STATE TRANSFER TABLE
EVRREC    DS     3000       *EVENT RECORD FILE
TODREC    DS     1000       *TIME OF DAY RECORD FILE
TIMREC    DS     1500       *DELAY TIMER RECORD FILE
RESREC    DS     1500       *RESPONSE RECORD FILE
DIGREC    DS     360        *DIGITAL OUTPUT RECORD FILE
*
*THE FOLLOWING MODULES REPRESENT THE MODULES
* REQUIRED FOR PROCESSING THE  OCCURRANCE OF
* A DIGITA EVENT.  THESE MODULES WOULD RESIDE IN PROM
*
*
* THE DIGITAL SCAN MODULE READS THE DIGITAL
* INPUTS. COMPARES THE CURRENT STATE WITH THE PREVIOUS
* STATE AND IF DIFFERENT INITALIZES THE PROCESSING
* OF THE STATE CHANGE.  UPON COMPLETION OF THE DIGITAL
* INPUT PROCESSING THE HISTORY STATE
* OF THE DIGITAL INPUTS IS UPDATED AND THE
* NEXT SCAN INITIATED
*
DIGSCN    IN     P1         *INPUT PORT 1
          STA    PORT1
          IN     P2         *INPUT PORT 2
          STA    PORT2
          IN     P3         *INPUT PORT 3
          STA    PORT3
          MVI    A,0        *RESET DIGITAL PROCESSING INDEX
          STA    DIGIND
*
* ISOLATE BIT FROM HISTORY AND CURRENT DIGITAL STATE
*
DIGI1     LDA    DIGIND     *GET PROCESSING INDEX
          CPI    8D         *INDEX .GT.8
          JC     DIG4       *JUMP IF INDEX=0-7
          CPI    16D        *INDEX.LT.16
          JC     DIG3       *JUMP IF INDEX =08-15
*
* PROCESS THIRD PORT
*
          LXI    H,PORT3    *HL=PORT3 DATA ADDRESS
          LXI    D,PORT3H   *DE=PORT3 HISTORY ADDRESS
DIG2      CALL   COMBIT     *GO COMPARE BIT
          CPI    0          *IF A=0 NO CHANGE
          JNZ    DIGPRO     *IF A=1 /R 2 GO PROCESS STATE CHANGE
DIG22     LDA    DIGIND     *INCREMENT PROCESS INDEX
          CPI    25D        *ALL INPUTS PROCESSED
          JNZ    DIGI1      *JUMP IF NOT
          JMP    DIGSCN     * GO RESTART SCAN
*
* PROCESS SECOND PORT
*
DIG3      LXI    H,PORT2    *HL=PORT 2 DATA ADDRESS
          LXI    D,PORT2H   *DE=PORT 2 HISTROY ADDRESS
          JMP    DIG2
*
* PROCESS FIRST PORT
*
DIG4      LXI    H,PORT1    *HL=PORT 1 DATA ADDRESS
          LXI    D,PORT1H   *DE=PORT1 HISTORY ADDRESS
          JMP    DIG2
*
* IF A= 1 PROCESS A SET STATE
* IF A=2 PROCESS A RESET STATE
* DIGIND CONTAINS THE RELATIVE INDEX INTO
* THE SENSOR STATE TRANSFER VECTOR TABLE
*
DIGPRO    STA    TEMP1
          LXI    H,XFVE     *HL=BASE ADDRESS
          LDA    DIGIND
          DCR    A
          RLC               *DIGIND*4
```

```
          LDA     TEMP2
          JMP     COM1
*
* THE EVENT RECORD PROCESSOR OBTAINS THE INDEX
* OF THE RESPONSE RECORD FROM THE EVENT RECORD
* CALCULATES ITS ADDRESS
* AND CALLS THE RESPONSE RECORD PROCESSOR
* EVENT RECORD ADDRESS IS IN H/L
EVPRO     MVI     A,ONE          *SET INITIAL INDEX
          PUSH    AF             *SAVE INDEX
          PUSH    HL
EVPR1     INX     HL             *INCREMENT RECORD ADDRESS
          PUSH    HL             *SAVE RECORD ADDRESS
          MOV     A,M            *A=RELATIVE RESPONSE RECORD INDEX
          CALL    RESPRO         *CALL RESPONSE RECORD PROCESSOR
          POP     HL             *GET EVENT RECORD ADDRESS
          POP     DE             *GET INITIAL RECORD ADDRESS
          POP     PSW            *A=GET INDEX
          INR     A              *INCREMENT INDEX
          XCHG                   *HL=INITIAL ADDRESS
          MOV     B,M            *B=NUMBER OF RECORDS TO PROCESS
          XCHG
          PUSH    DE             *SAVE INITIAL RECORD ADDRESS
          CMP     B              *HAVE ALL RESPONSE RECORDS BEEN PROCESSED
          JZ      EVPR2          *JUMP IF MORE TO PROCESS
          POP     HL             *REMOVE BASE ADDRESS OF RECORD FROM STACK
          RET
EVPR2     POP     DE
          INR     A
          PUSH    PSW
          PUSH    DE
          JP      EVPR1          *GO PROCESS NEXT ENTRY IN RECORD
*
* THE RESPONSE RECORD PROCESSOR OBTAINS THE
* RESPONSE TYPE FROM THE RECORD AND TRANSFERS CONTROL
* TO THE APPROPRIATE RESPONSE MODULE
* ON ENTRY A=RESPONSE RECORD INDEX
*
RESPRO    MOV     L,A
          MVI     H,ZERO         *HL=RESPONSE INDEX
          MOV     B,H
          MOV     C,L
          DAD     DE
          DAD     DE
          DAD     DE             *HL=RESPONSE INDEX*3
          XCHG
          LXI     HL,RESREC
          DAD     DE             *HL=RESPONSE RECORD ADDRESS
          MOV     A,M            *A=RESPONSE TYPE
          ADD     A
          INX     HL             *HL=ADDRESS OF NEXT DATA ITEM IN RECORD
          XCHG                   *DE=ADDRESS OF NEXT DATA ITEM IN RECORD
          MOV     C,A
          MVI     B,ZERO
          LXI     HL,RESTAB
          DAD     BC             *HL=TABLE ADDRESS
          MOV     C,HL
          INX
          MOV     H,HL           *BC=ADDRESS OF RESPONSE MODULE
          PUSH    BC
          XCHG
          INX     HL
          MOV     F,M
          RET                    *GO TO RESPONSE SUBPROCESSOR
RESTAB    DW      DIGOUT
          DW      APPL
          DW      DELINI
          DW      DELDAC
          DW      TODACT
          DW      TODDAC
*
* THE DIGITAL OUTPUT SUBPROCESSOR EXTRACTS
* THE PORT/MASK/AND DIRECTIVE INFORMATION FROM
* THE INDICATED RECORD AND OUTPUTS THE NEW VALUE
* DE CONTAINS THE RECORD INDEX
*
DIGOUT    LXI     HL,DIGREC      *SET DIGITAL OUTPUT RECORD ADDRESS
          DAD     DE
          DAD     DE
          DAD     DE             *HL=OUTPUT RECORD ADDRESS
          PUSH    HL             *SAVE OUTPUT RECORD ADDRESS
          MOV     A,M            *A=PORT
          DCR     A
          MOV     C,A
          MVI     B,ZERO
          LXI     HL,OPORT       *HL=OUTPUT PORT BASE ADDRESS
          DAD     BC             *HL=ADDRESS OF PORT DATA
          MOV     A,M            *A=OUTPUT PORT CURRENT STATE
          POP     HL
          INX     HL             *HL=MASK ADDRESS
```

```
          RLC
          MOV     C,A
          MVI     B,ZERO
          DAD HL,BC
          LDA     TEMP1
          CPI     2
          JZ      DIGP1           *JUMP IF SET TO RESET
          INX     H
          INX     H
DIGP1     LDAX    DE              *DE≡TRANSFER VECTOR
          CPI     80H             *IS HIGH ORDER BIT SET
          JNC     DIG22           *JUMP IF RECORD INACTIVE
          LDAX    DE
          MOV     L,A
          INX     DE
          LDAX    DE
          MOV     H,A
          CALL EVPRO              *PROCESS EVENT RECORD
          JMP     DIG22
*
* COMBIT ISOLATES AND COMPARES THE DATUM BIT
* IN THE CURRENT AND HISTORICAL DIGITAL INPUT
* STATE TABLES
*
* HL≡CURRENT DATA ADDRESS
* DE≡HISTORY DATA ADDRESS
*
COMBIT    LDAX    D               *A≡HISTORY DATA
          CMP     M               *COMPARE HISTORY TO CURRENT
          JZ      COM5            *JUMP IF CURRENT≡HISTORY
*
* NO COMPARE   ISOLATE BIT
*
          MVI     A,1             *SET MASK≡1
          JMP     COM2
COM1      CPI     0               *SHIGT COUNT≡0
          JZ      COM4            *JUMP IF SHIFT COUNT≡0
          RLC     A               *SHIFT MASK BIT
COM2      STA     TEMP2           *SAVE MASK
*
* GET CURRENT STATE BIT
*
          MOV     B,M             *B≡CURRENT DATA
          ANA     B               *A≡CURRENT STATUS BIT
          STA     TEMP3           *TEMP3≡CURRENT BIT STATUS
          XCHG
          MOV     B,M
          LDA     TEMP2
          ANA     B
          MOV     B,A             *B≡HISTORY BIT
          LDA     TEMP3           *A≡CURRENT BIT
          CMP     B               *COMPARE BITS
          JZ      COM6            *RETURN IF EQUAL
*
* CURRENT .NF. HISTORY
* UPDATE HISTORY STATE
*
          CPI     0               *IS BIT RESET
          JNZ     COM3            *JUMP IF BIT SET
* RESET HISTROY STATE
          LDA     TEMP2           *GET MASK
          CMA                     *COMPLIMENT MASK
          XCHG
          ANA     M
          MOV     M,A             *STORE HISTORY
          INR     A
          STA     DIGIND
          MVI     A,2
          RET
*
* SET HISTORY STATE
*
COM3      LDA     TEMP2           *GET MASK
          ORA     M               *A≡NEW HISTROY
          MOV     M,A             *STROE NEW HISTORY
          LDA     DIGIND
          INR     A
          STA     DIGIND
          MVI     A,1
          RET
*
COM5      LDA     DIGIND
          ADD     8D
          STA     DIGIND
COM4      MVI     A,0             *A≡0
          RET
COM6      XCHG
          LDA     DIGIND
          INR     A
          STA     DIGIND
```

```
              ANA       M
              PUSH      PSW             *SAVE MASKED STATE
              INX       HL
              MOV       A,M             *GET SET/RESET DIRECTIVE
              DCX       HL
              CPI       ONE
              JZ        DIGSET          *GO TO SET FUNCTION
              POP       BC              *B=NEW OUTPUT STATE
DIGOU1        DCX       HL              *HL=PORT ADDRESS
              MOV       A,M             *A=OUTPUT PORT
              STA       DIGOUT2+1       *SAVE PORT NUMBER IN OUTPUT INSTRUCTION
              MOV       A,B             A=OUPUT DATA
DIGOU2        OUT       THREE
              MOV       C,M             *C=PORT
              MOV       B,ZERO
              LXI       HL,OPORT
              DAD       BC
              MOV       M,A             *SAVE NEW STATE IN MEMORY
              RET
DIGSET        MOV       A,M             *A=MASK
              CMA
              POP       BC              *C=MASKED VALUE
              ORA       B               *A=NEW OUTPUT STATE
              MOV       B,A
              JP        DIGOUT1
*
* VECTORS TO APPLICATION TASK
* DE CONTAINS MODULE ADDRESS
*
APPL          PUSH      DE              *SAVE ADDRESS ON STACK
              RET                       *EXIT TO APPLICATION MODULE
*
* DELINT ACTIVATES THE DELAY RECORD WHOSE
* RECORD INDEX IS IN DE
*
DELINT        LXI       HL,DELREC       *SET BASE ADDRESS
              DAD       DE
              DAD       DE
              DAD       DE
              DAD       DE
              DAD       DE
              DAD       DE              *HL=DELAY RECORD ADDRESS
              MOV       A,M             *GET ACTIVE FLAG
              ANI       80H             *IS ACTIVE FLAG RESET ,ACTIVE<
              RZ                        *RETURN IF RECORD IS ACTIVE %RESET<
              INX       HL
              INX       HL              *HL=TIMER ACTIVATION VALUE
              MOV       A,M             *A=TIMER VALUE
              DCX       HL
              DCX       HL
              MOV       M,A             *SET VALUE IN RECORD
              RET
*
* DELDAC DEACTIVATES THE DELAY RECORD
* WHOSE INDEX IS N DE
*
DELDAC        LXI       HL,DELRES       *SET BASE ADDRESS
              DAD       DE
              DAD       DE
              DAD       DE
              DAD       DE
              DAD       DE
              DAD       DE
              MVI       A,ZERO
              MOV       M,A             *RESET ACTIVE FLAG
              RET
*
* TODACT ACTIVATES THE TOD RECORD
* WHOSE RECORD INDEX IS IN DE
*
TODACT        LXI       HL,TODREC       *SET BASE ADDRESS
              DAD       DE
              DAD       DE
              DAD       DE
              DAD       DE
              DAD       DE
              MVI       A,7FH           *ACTIVATION MASK
              ANA       M               *SET RECORD ACTIVE
              MOV       M,A
              RET
TODDAC        LXI       HL,TODREC       *SET BASE ADDRESS
              DAD       DE
              DAD       DE
              DAD       DE
              DAD       DE
              MVI       A,80H           *DEACTIVATION MASK
              ORA       M               *SET RECORD INACTIVE
              MOV       M,A
              RET
```

This time it determines that the response required from record 1 is to call the application module located at hexadecimal address 2C0F. This information is transmitted to APPL (the application task initiator subprocessor) by the response processor, which transfers control. In our particular example this application module will cause the display panel to flash the location of the sensor giving the alarm. Upon completion of this function the application module returns control to the event record processor.

Next, the event record processor transfers the index of response record 2 to the response processor. Recognizing that a digital output is to be initiated, the processor extracts the index of DIRREC (the digital output record), 0 in this example, and initiates DIGOUT (the digital output subprocessor).

The function of this subprocessor is to activate the audible alarm. A 3 byte record is used to effect the actual output:

Digital Output Record

Byte 0 = output port;
Byte 1 = bit isolation mask;
Byte 2 = 0 for reset (off);
       1 for set (on).

In our case let's assume that the audible alarm has been connected to bit 5 on output port 3 and must be set to a logic 1 to sound the alarm. This requires a digital output record as follows:

Audible Alarm Output Record

Byte 0 = 3;
Byte 1 = 20;
Byte 2 = 1.

Using this information bit 5 on output port 3 is set and the alarm horn is turned on. Processing this final response causes the event processor to return to the digital scan module, which will continue monitoring the state of all the digital inputs.

The intent of this example is to show how the system works when attached to the real world. While this example does not cover all the functions one might associate with a smoke or fire alarm, it does serve to illustrate how the various tables are structured. What you can do with such a system is limited only by your imagination. As stated earlier in this series, the system can be structured to perform many of the discrete tasks associated with the control and monitoring of the home or office, as well as protecting your property against intruders. So let your imagination take command and have fun. ∎

ULTRASONIC TRANDUCERS LOCATED

Dear Steve,

Just a short note to ask for a helping hand and to let you know that your articles in *BYTE* are read and enjoyed.

In your article "Building a Computer Controlled Security System for Your Home," you showed a circuit for generation and detection of ultrasonic signals. Would you be so kind as to note a company or two that I could contact for transducers? I have an application for ultrasonics and need a little help in knowing where to write.

Thank you for your help, and again, I enjoy your articles.

Tom Yocom

The particular transducers used in the article are from MASSA in Hingham, MA. I obtained them through Bullet Electronics, POB 401244E Garland TX 75040, (214) 278-3553. I suggest calling them to determine price and availability. Since I usually purchase components in large quantities long before I actually need them for an article, I hesitate to quote a price and a definite source. The MASSA units had an output frequency of 23 kHz....Steve

Dear Steve,

While sitting in my living room last summer watching Hurricane David whirl by, I wanted nothing more than to use my TRS-80 computer. Unfortunately, our power was out for several hours, and when it came back on, my work was complicated by several brief power interruptions. Has anyone developed a combination emergency and uninterruptible power supply suitable for home-computer systems?

My approach to this problem would start with a well-shielded transformer and regulated battery charger. A zener regulator would float-charge a sealed maintenance-free automobile battery at the manufacturer's recommended voltage to ensure long life. Rather than use a square-wave-type inverter, a crystal-controlled 60 Hz oscillator might be more appropriate, driving a 250 W amplifier that would produce a reasonable approximation of standard AC power. This would provide electricity for my computer and several peripheral devices, including a light bulb.

R B Nottingham

I have been thinking about uninterruptible power quite a bit lately. I first mentioned it in my articles on computer-controlled security for the home in the January through March 1979 issues of BYTE.

I hesitate to guess at the cost of a 250 W amplifier with a peak output voltage of 176 V. In my own system I have battery backup sufficient for a half hour. The battery is connected directly to the power-supply regulators, and the system shuts down automatically before the power runs out.

The dilemma I face is that everything in my house is electronically controlled, even the wood stove. (See "A Computer-Controlled Wood Stove," Chapter 15.) My uninterruptible house requires that I walk out to the garage and start my 5 kW propane-fueled generator, while the computer is running under battery power....Steve

MORE POWER

Dear Steve,

I noticed your comment on UPSs (uninterruptible power supplies), and thought I would mention that they are commercially available in sizes small enough to be useful to personal-computer users (see the Hardside catalog, page 34). I do not know who the actual manufacturer is, but I would like to know more about these items. The devices I am concerned with have specifications that accommodate 60 and 120 Hz power, with and without surge protection, and supply 150 or 200 W. The trade name is "Mayday."

R M Sanford

Thank you for pointing out the Mayday UPS. It is manufactured by Sun-Technology Inc, which is located in New Durham, New Hampshire. The Mayday UPS is available from Hardside, 6 South St, Milford NH 03055, (800) 258-1790. According to the Hardside catalog, prices begin at $168....Steve

Ciarcia's Circuit Cellar

The Toy Store Begins at Home

"Mister? Mister?"

A little boy was tugging on my sleeve. It startled me that in today's sophisticated society anyone would attempt to attract my attention by such an obvious, though effective, means. Impatient and undaunted by the scowl I flashed in his direction, he said, "Mister? Do you know where the toy department is?"

I have never acquired what some people call the ability to commune with children. Perplexed therefore as to the presentation of a proper reply, I considered an indignant, wave-of-the-hand dismissal of "Over there, kid." On the other hand, should I consider a character reversal with a Santa Claus imitation and invite the young man to hop up on my shoulder while we looked over the store directory together? The latter seemed hardly my style and the former was much too harsh even considering his still firm attachment to my sleeve.

"Mister? Mister?"

The delay only heightened his fervor.

I looked up and found myself staring straight at the shirt pocket button of a very large man. Instantly I calculated that this male figure dressed in jeans, heavy boots and a woolen shirt was a foot taller than I. His relationship with the boy was quickly clarified as he said in a deep paternal voice, "Come on Brucie, I think it's over there where that crowd is." I waited for Paul Bunyan and son to be safely on their way before I made my next move.

Stark reality returned, however, when I remembered that I, too, was looking for the toy department. It verges on humiliation actually. Why do they have to categorize everything? Just because an item is manufactured by a toy company doesn't immediately classify it as a toy. I mean, big people have constructive leisure time manipulatives and little people have toys. Department stores should realize the embarrassment of crossing this line and have an "amusements for the sophisticated" department and a "toys for tots" department.

Finding the toy department was no prob-lem. I simply stood where I was and slowly rotated 360°. The noise peaked at about 160° SSE and I cautiously proceeded in that direction. The noise in my immediate vicinity became sharply amplified as two young boys raced by, carrying some unidentifiable toy devices.

I spied my objective ahead — the electronic games counter. I got into line between two youngsters and their parents. Were these PG or R rated games? I saw no parents with the kids playing basketball in the next aisle. Perhaps the cost of computerized games warranted closer parental scrutiny. $5 for a hockey stick is one thing, but $50 for a talking plastic robot is another. All the games at this counter incorporated microprocessors as their intelligence. Some simulated war games and produced authentic battle sounds while others proved to be formidable challengers in games of chance.

I looked through the products in the case, hoping to spot the one I so desperately wanted. Would this be another store that was completely sold out? Would I never get my Simon?

"Sir? Can I help you?" the salesman asked. His attitude was surprisingly pleasant considering that he worked in the store's combat zone.

"I don't see it!"

"See what, sir?"

Still vaguely pleasant, his tone changed to "I've had a long day, buddy. Let's not play 20 questions."

"Simon, of course!" I replied. "But I know you don't have any. No one does."

"You're in luck, sir. I believe we received a back-ordered shipment yesterday. I'll check."

A young girl behind me said, "Did you hear that, mommy? They have Simon! I can practice for the competition after all."

I said, "Competition? Simon?"

"Sure. Everybody's got one. Except me, that is. We have contests in school to see who can remember the longest tune. It's fun. Oh, I can't wait!" she responded, tugging on my sleeve.

Simon is a trademark of the Milton Bradley Corporation.

"That sounds exciting. I hope you do well in the contest," I said.

The salesman returned.

"I have one left. You're in luck."

I hardly had time to smile as he passed it to me. I heard a whimper from behind me and sensed the little girl's disappointment.

Saying nothing I turned to look at her. She tried to hide her anguish.

"What is your name, little girl?" I asked, stooping down a bit to be more at her level.

"Brenda," she said wistfully.

"That's a coincidence. I have a little...er, girl named Brenda too." I had to catch

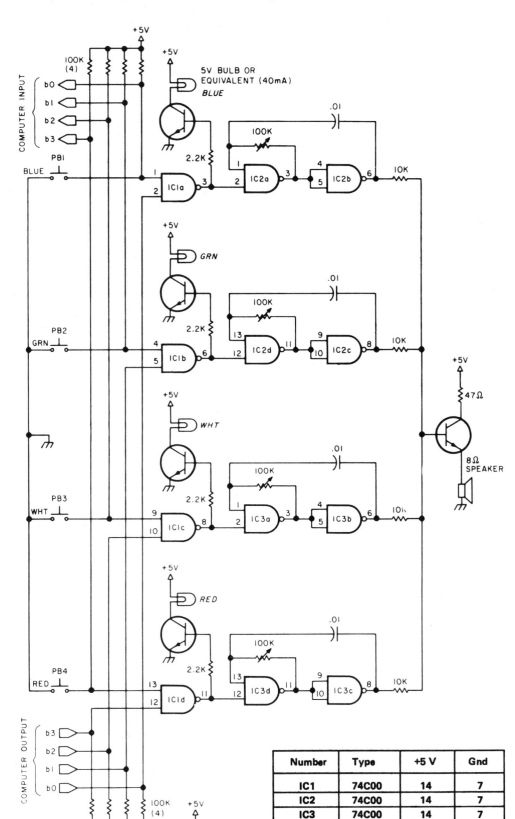

Figure 1a: Hardware tone generator for the musical tone sequencer. The computer plays a sequence of lights and associated tones and detects the player's response. (All transistors are 2N2222.)

Number	Type	+5 V	Gnd
IC1	74C00	14	7
IC2	74C00	14	7
IC3	74C00	14	7

myself — as I have a female Scottish Terrier named Brenda. Parents might get upset if you compare their children to dogs. "She's a little smaller than you are."

"Is Simon for her, Mister?"

"No, she likes playing with tennis balls. But no matter. I've only been *looking* at this game. I'm not sure I really want to buy it just yet. Would you like it?"

She offered several relieved thank-you's as I bolted for the door. I was in a hurry to get to the department store two blocks up the street before they closed. . . .

Musical Games Are Addicting

Some time ago I was in a stuffy business meeting. When it became apparent to the chairman that most of the attendees were asleep, he pulled out a saucer shaped object with four colored areas on it and slid it along the table. It stopped in front of me and went "beep" and lit a red light. Instructed to respond in kind, I pressed the red area which turned out to be an oversized lighted pushbutton. The saucer replied "beep-boop" and lit the red and green lights sequentially. It became immediately apparent that the plastic saucer was a game and the object was to duplicate the sequential tones it played. The task became increasingly difficult as it added another note each time around. If missed, it made a sound like a "raspberry" before starting a new game.

This "game" turned out to be Simon, from Milton Bradley Corporation. It uses a microprocessor to synthesize the tones, light the lights, and generate the sequence.

Build Your Own Musical Game

It is only logical that any of the $30 to

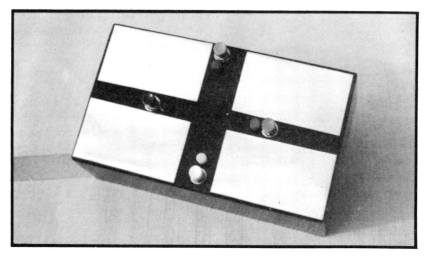

Photo 1: *Player console for the computerized musical tone game. Players attempt to repeat a sequence of tones and corresponding lights chosen by the computer at random.*

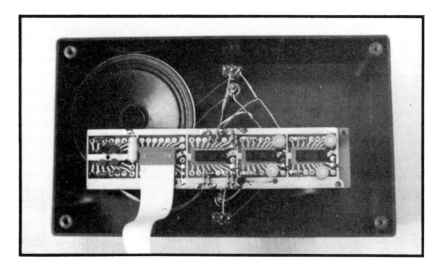

Photo 2: *Bottom view of the player's console. The ribbon connector attaches to the user's personal computer.*

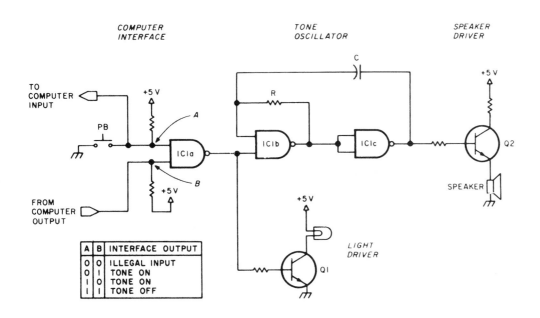

A	B	INTERFACE OUTPUT
0	0	ILLEGAL INPUT
0	1	TONE ON
1	0	TONE ON
1	1	TONE OFF

Figure 1b: *Details of the circuit in figure 1a, showing one of the four light and sound generating sections.*

$50 electronic toys in department stores can be simulated with the average $6000 personal computer. (This is why critics frequently call computers illogical.) The distinguishing feature between a toy built around a microprocessor and the average home computer is the packaging and I/O (input/output) interface. With the exception of addressable memory, the microprocessor in a battleship game has a processing capability comparable to the more general purpose processors like the 8080 and 6800. The major difference is that single chip computers incorporate limited quantities of programmable memory, read-only memory, and I/O in one package. This is the most cost-effective approach for a dedicated task like a game. The most popular single chip computer in the computer games market is the Texas Instruments TMS 1000. Customized versions of this integrated circuit are used in the majority of electronic games.

Presuming that we can write a program on our large computer that accomplishes

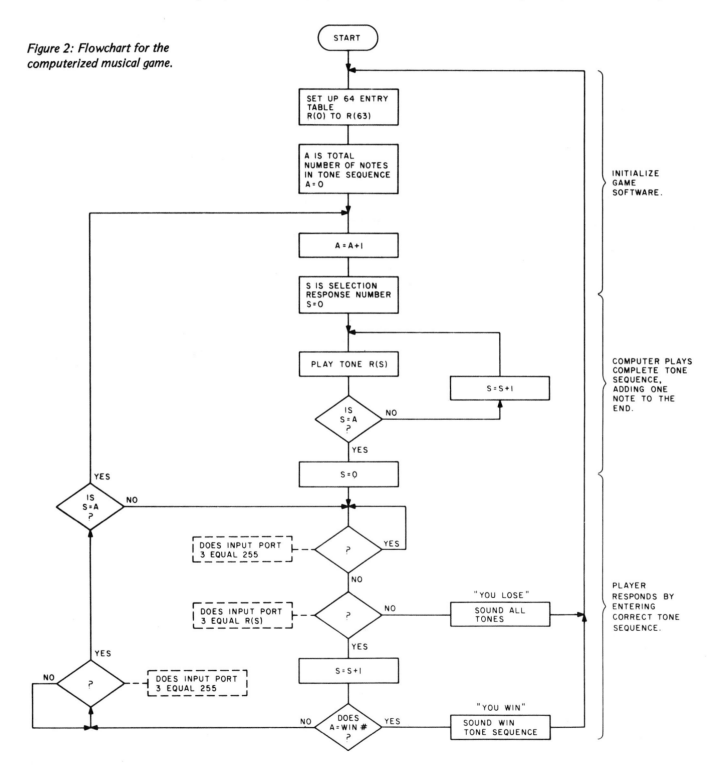

Figure 2: Flowchart for the computerized musical game.

the same logical objective as the dedicated game, the only real difference becomes I/O. Most personal computers incorporate ASCII keyboards, video displays, and cassette-tape interfaces for I/O. Electronic board games use a few switch inputs (constant closures) and lights or buzzers for output and, because there is little operating system overhead, sound effects are directly synthesized by program timing loops. Theoretically, if we attach these switches and lights to a convenient I/O port on our computer we should be able to program a similar or even more challenging game.

Building a musical game that tests the players' ability to memorize a string of tones is a simple task. Input to the computer consists of four switches, one for each of four tones. Output from the computer is likewise four signals which light four colored lights on the player console. Each light corresponds to a distinctive tone.

The game is simple to play. The computer plays a tone and the player responds by pressing the button for that same tone. Next, the computer plays two notes and the player replies accordingly. Each correct exchange results in adding one more note to the string. Eventually either the player misses by being unable to replay the exact tone sequence, or wins by attaining some preset number of notes without failure. The former is signified by an ungracious combination of tones and the latter by a distinctive tune played by the computer in celebration.

There are two possible design approaches. One is to use machine language and a "bare bones" interface consisting of four switches and four lights directly connected to a parallel input and output port. Timing loops written into the software produce the tones. This method uses the least hardware but requires considerably more software.

The second alternative is to use a high level language such as BASIC and use an external hardware interface for tone generation. This is the approach I have taken. Experimenters wishing to use another approach can easily follow the logic flow of BASIC and in this way I am not confining the reader to a particular microprocessor. Also, on-the-spot program variations to accommodate individual players are more easily implemented in a high level language.

Figure 1a illustrates the hardware interface of this musical game; photos 1 and 2 demonstrate typical layouts. A more detailed description of an individual tone generating section is given in figure 1b. Normally, both signal points A and B are at a high logic level and the tone is off. The tone and light can be turned on by either a low output signal from the computer or

the pushbutton being pressed. The resulting high level output of IC1a turns on the oscillator formed from IC1b and IC1c and drives the light through transistor Q_1.

A flowchart of the software as written in BASIC is shown in figure 2. When the game is initialized, a random number generator sets up a tone sequence of 64 notes. After playing the first note it waits for the player's response and then repeats the action adding another note. The software is written so that the speed of player response is not important. Player frustration is strictly limited to remembering the tone sequence. The BASIC program which plays this game is shown in listing 1.

I have found that this game is a good way to demonstrate my computer to people

Listing 1: Program for the musical-tone game, written in 8 K Zapple BASIC.

```
 90  REM
 92  REM ** CIARCIA'S CIRCUIT CELLAR COPYRIGHT 1979 **
 94  REM
100  PRINT"THIS IS A MUSICAL GAME TO TEST YOUR MEMORY"
105  REM
110  REM
115  REM FIRST THING WE DO IS SET UP A TABLE OF 64
120  REM RANDOM NUMBERS WITHIN THE CHOICES OF 1,2,4, OR 8.
125  REM THESE NUMBERS ARE SINGLE BITS WHICH INDICATE A
130  REM PARTICULAR TONE AND COLORED LIGHT.
135  REM THE COMPUTER INTERFACE IS BITS 0 THRU 3 OF I/O PORT 3
140  REM
200  DIM R(64) :DIM S(64) :DIM A(64)
205  A=0
210  FOR S=0 TO 63
220  R=INT(RND(1)*10)
230  IF R>3 THEN 220
240  R(S)=255-2^R :REM THE INPUT TO THE INTERFACE IS LOW TRUE LOGIC
245  REM TO TURN ON A TONE ALL BITS ARE HIGH EXCEPT THE
247  REM ONE WHICH IS TO BE COMMUNICATED
250  NEXT S
260  REM
270  REM
400  S=0 :A=A+1
410  OUT 3,R(S) :GOSUB 2000 :REM TURN ON TONE
420  OUT 3,255             :REM TURN OFF TONE
425  S=S+1
430  IF S=A THEN 450 ELSE 410
450  S=0
460  W=INP(3)
465  IF W<>255 THEN 470 ELSE 460:REM HAS A BUTTON BEEN PUSHED?
470  IF W=R(S) THEN 480 ELSE 600
480  S=S+1
481  REM A IS PRESET TO EQUAL WIN NUMBER. THIS CAN BE 1 TO 64 TONES
482  IF A=16 THEN PRINT"YOU WIN":GOTO 700
490  W=INP(3)
495  IF W<>255 THEN 490 :REM HAS THE PLAYER RELEASED THE BUTTON?
500  IF S=A THEN 580
510  GOTO 460
520  REM
530  REM
580  REM RETRY DELAY
585  FOR T=0 TO 3 :GOSUB 2000 :NEXT T
590  GOTO 400
600  PRINT"SORRY,YOU MISSED IT . . . . . YOU HAD ";A;" NOTES IN THE
     SEQUENCE"
605  PRINT"TRY AGAIN"
610  OUT 3,0 :REM TURN ON ALL TONES
620  FOR T=0 TO 3 :GOSUB 2000 :NEXT T
625  OUT 3,255
630  GOTO 205
700  FOR T=0 TO 6 :REM PLAY TUNE TO INDICATE A WINNER
705  OUT 3,254 :GOSUB 2050 :OUT 3,253 :GOSUB 2050
710  OUT 3,251 :GOSUB 2050 :OUT 3,247 :GOSUB 2050
715  OUT 3,255 : NEXT T
720  GOTO 205
1980 REM
1990 REM THE VALUE OF T1 SETS THE TONE DURATION
2000 FOR T1=0 TO 250 :NEXT T1 :RETURN
2050 REM WIN DELAY TIMER
2060 FOR Q1=0 TO 80 :NEXT Q1 :RETURN
```

Photo 3: Corner of the circuit cellar showing the 64 K dual floppy disk Z-80 system used by the author to drive the musical tone game.

totally unfamiliar with them. Some of my more computer oriented friends jokingly suggest that I may be doing things the hard way using a 64 K byte dual disk Z-80 system for the game.■

Ciarcia's Circuit Cellar

Photo 1: Example of an optical fiber transmitting a very bright light. The conductor is a single 40 mil plastic fiber. The light is generated by a helium-neon laser.

Communicate on a Light Beam

Coming up out of the Circuit Cellar is a rare occurrence, to the point where some of my friends have accused me of being a mushroom. I prefer to be likened to a mole—a more dignified species. We share a common bond of subterranean existence and fear of bright sunlight, but the mole's predicament is dictated by nature, and mine by choice.

The Circuit Cellar is by no means a hole in the ground. It's heated, well-lit and looks more like a living room than a cellar. Even though it affords all the comforts of home, there are those occasions when a change of environment is required. It's not enough to walk out in the driveway, take a deep breath and run back into the cellar. Sometimes a complete change of surroundings is needed to shock the mind out of the doldrums and spark creativity (eg: a vacation). Since I usually don't have time for vacations, I take "business excursions for purposes of cerebral detoxification" or "ECDs" for short.

For two months I had been wrestling with the details of an article on fiber optics and laser communications (this one). The hardware was completed very quickly, as with most of my projects, but the text dragged on for weeks. Lighting the wood stove in the Circuit Cellar became an all too easy chore using the piles of scrap paper I was generating. My graphospasms (ie: writer's cramps) were not bearing fruit. One time I even found myself sitting at my desk pushing pencils through the electric pencil sharpener until it started smoking.

During times like this there was only one place to go — New Hampshire — to see the Colonel. My father-in-law, Colonel Foster, was the one person who could break me out of this slump. Between stories about old army buddies and spending the war in the Aleutians waiting for an invasion I would surely find some inspiration.

"Colonel? Are you there?" After anxiously dialing his telephone number and saying hello, I was left with silence at the other end of the line. . .

"Colonel?"

"Be right with you, Steve." As the receiver was picked up again he apologized, "Sorry Steve, my man was at bat and I had to see the hit. You're a Red Sox fan, aren't you?"

It would be in bad taste for me to suggest that my subterranean hideaway provided all the spiritual stimulation I needed and that chasing a little ball around in the grass was not in my spectrum of pursuits.

"I quite understand your enjoyment of the game, Colonel. I hope your team wins," I replied, evading his question. During my statement I heard him roar again in response to the activities on the television. When I sensed a lull, possibly precipitated by a commercial, I continued, "Colonel, I need to get away. How would you like some company tonight?"

"Sure, you know you're always welcome. I haven't had anyone to tell a good army story to in a long time."

I told him I'd pack all the gear in the car and be there in three hours. Possibly I would feel better about writing once I arrived.

The Colonel, sensing the termination of the commercial, quickly responded, "Three hours is great. The game is still in the first inning. If you hurry you may get here before it's over. . .gotta go now."

One of the good things about living in New England is that everything is close. It was a scant 3 hour drive between Connecticut and New Hampshire, but I dragged it out an extra half hour so I wouldn't be competing with the Red Sox for the Colonel's attention. As I pulled into the garage he came out to greet me.

"Howdy," he said, slapping me on the back. From his exuberance I could tell that the Red Sox had just won the game.

"Come on in and get settled. I'm expecting a telephone call. . .oops, there it is now."

Leaving the electronics junk in the car I followed him into the house. He was still wearing his lucky Red Sox baseball cap as he spoke.

"Chester, wasn't the game great? I thought they were going to blow it in the 6th. . .You bet, I'm ready for tomorrow's game. If they can play like that again, the pennant is in the bag. . ."

Suddenly Colonel Foster's expression changed to amazement, then anger. He grabbed his cap, slung it into the chair he was standing near and complained, "Darn woman again!. . .What do you mean lucky! The Red Sox won through skill, not luck!. . .Go play with your WATS lines and let Chester and me talk." It was obvious that suddenly there was a third party to their conversation.

"Beatrice, I don't care if you think it was an error. It was ruled as a single!. . . Yes, I know the 6th looked bad but that still doesn't mean they're just lucky. . ."

It was becoming an argument between the Colonel and Beatrice. A hint as to her identity was provided when he responded, "Beatrice, would you keep your opinions to yourself and let me talk to Chester? Chester, come on over for a private talk!"

He slammed the reciever down on the phone, put his baseball cap back on, and slumped into the easy chair. "I just can't carry on a baseball conversation with that woman around."

"Who's Beatrice?"

"The switchboard operator for the town. We don't have all that new computer telephone stuff you city slickers have. We have Beatrice. When it's business or personal she's good and keeps her nose out. But,

when it's baseball, Beatrice has to get her two cents in!"

(Obviously what the Colonel and Chester needed was an alternate means of communication, such as CB.)

"I've got a great idea, Colonel. Why don't you and Chester use CB radios instead of the telephone?" The Colonel led me to the bookcase in the study. I found myself staring directly at a CB radio. He flipped it on and said, "Tune in channel 19 and listen." The radio came to life. "Breaker one nine. . .breaker one nine. . . this is your Big Mama on this one niner. . . all you 18 wheelers just put the hammer to the floor and let Big Mama be your guide. . .I'll have a Smokey report in five, but first, the weather. . ."

My eyes opened wide. "Is that Bea. . ."

"Beatrice? You're darn tootin' it is. She's got an antenna tower on her house and radio gear that would put an FCC test laboratory to shame. I swear she's running a full gallon."

"We tried CB a while back and it was useless." This time the conversation came from behind. Chester had let himself in and joined us in the study. He continued, "It all started when we telephoned the games to the tower."

"Tower?"

"I'm sorry, I guess the Colonel didn't tell you." Walking over to the window of the study and pointing to the mountain top roughly two miles away, he said, "You see that structure on top of that hill? That's my tower. Well, not exactly *my* tower. I just work there. It's a combination fire tower and radio relay station. Occasionally I have to sit up there and monitor equipment during important transmissions."

"What's that got to do with Beatrice?"

"With all the interference from the equipment up there I can't use a radio or television to watch the Red Sox."

(This was beginning to take on the aspects of a good mystery.)

"The Colonel would tune in the game on his television set here, telephone me in the tower and then lay the receiver near the television so I could listen to the game. When Beatrice found out, she'd bust in and add her commentary to the game. Do you know what it's like having a nosey Howard Cosell-type beating on your ear for three hours at a time?"

I could only offer my sympathy. If there were a solution short of stringing two miles of wire I didn't see it yet. But I would continue to think about it.

"Tomorrow is a very important Red Sox game. The pennant may hinge on it. Unfortunately, tomorrow is also a day I have to spend in the tower. I really want to listen to the game, but Beatrice is tough to listen to."

I ran over to the window, looked at the tower in the distance, and noted the glass windows circling the observation deck. "What's the weather report for tomorrow?"

"Cloudy and cool, I think," Chester answered.

"Good! Clear weather. . .Colonel, could the television set be moved in this room for the game tomorrow?"

"I suppose so. Why?"

I scanned the study looking for a convenient AC power outlet and spied one by the window.

"Perfect," I said.

Both the Colonel and Chester were a little perplexed at my behavior.

"What if I told you there were a way for Chester to listen to tomorrow's game undisturbed by Beatrice?"

"We've tried everything. What are you planning?"

"Wait here and I'll show you." I dashed off to my car and took a tripod, a long white rectangular instrument, a small black box with a lens at one end and a few patch cords out of the trunk. Dragging all the equipment into the study, I proceeded to assemble it, much to their amazement.

"What's all this, Steve?" the Colonel asked.

With as straight a face as I could muster I replied, "It's a laser."

Both men, army veterans of two wars and thirty years' service, took two steps back and exclaimed, "A laser?" It was instantly apparent that the words "laser" and "death ray" were synonymous for them. Before I let them think I planned to rub out Beatrice, I quickly continued my explanation.

"There are big lasers and little lasers. This is a little one. It won't burn anything or hurt anyone if used properly. Eye protection is the only consideration necessary on this particular laser."

"Do you always carry this stuff around with you?" the Colonel asked.

"No. It just happens to be the topic of this month's article for *BYTE*."

"What has this got to do with tomorrow's game?" Chester asked.

"We're going to transmit the game to you in the tower on a beam of light."

Their eyes opened wider but they remained receptive.

"Let me demonstrate."

I took the transistor radio, tuned it to a station and placed it on the coffee table. Taking a long patch cord, I plugged one end in the radio earphone jack, automatically silencing the radio speaker, and plugged the other into the rear of the laser. Aiming the laser, I turned it on. A red spot, about 1/8 inch diameter, shone brightly on the wall 15 feet away.

"You're sure that won't burn the wall?"

"Trust me."

Next, I picked up the black box with the lens on it and turned it on. I walked over to the illuminated spot on the wall and interrupted the laser beam path with the box. When the beam intersected with the lens, music was heard!

"That's the radio station you tuned in, all right," Chester said.

"Colonel, take that poker from the fireplace and wave it back and forth in front of the laser so it interrupts the beam."

"Why. . .the radio goes on and off," he exclaimed a minute later.

"Correction, Colonel. The radio doesn't go off, only the receiver, when it no longer 'sees' the modulated laser light beam. Notice in addition that the beam barely spreads out at all over the 15 feet to the wall."

"I think I get what you're driving at, Steve."

"You've got it. Chester takes the receiver up to the tower tomorrow, aims it at this window using the gun sight scope on top. Then we turn on the laser which, instead of being connected to the radio, comes from the television. Voila! Instant uninterrupted Red Sox baseball. And, no Beatrice!"

"Will it really work, Steve?" Chester asked.

"Sure, and tomorrow we'll prove it."

Before the next comment from anyone, the telephone rang and Colonel Foster answered it. Chester and I listened and smiled.

"Look, Beatrice, your team doesn't have a chance for the pennant. . .Are you still claiming that that was an error?. . .It wasn't just luck in the 6th I tell you. . ."

Chester and I laughed. Beatrice was really giving the Colonel a run for his money, but there was a twinkle in his eye as he spoke. The Colonel was living what he enjoyed most — baseball. First on television and then blow by blow with Beatrice.

Communicate on a Light Beam

Most experimenters have never considered using a modulated light beam for data communication. I'm not suggesting that everyone throw out their twisted pair RS-232 lines and replace them with laser beams, but I do ask you to consider the commercial advantages of such a concept and try a few experiments.

When discussing modulated light communications, a definition of terms is in order. The two most often heard are lasers and fiber optics. It is important to recognize that one is a light source and the other is a light conductor. It is not necessary for them to be used together but this is often the case. I'll explain more about each later.

A full duplex optical communication link is shown schematically in figure 1. It consists of two pairs of optical transmitters and receivers which allow data to flow in two directions simultaneously. Data from the base to the remote travels on one line, while data from the remote to the base is on the other. This is a dedicated duplex hookup. Unlike the ones you've probably used, this one uses fiber optic cable rather than wire. In its commercial applications it can offer the following advantages:

- Immunity to strong electrical or magnetic noise. Fiber optic material is usually glass or plastic and since there is no electrical conduction there can be no induced electrical noise.
- High electrical isolation. Since the data conductor is a dielectric material, the isolation between the transmitter and receiver is a function of distance.
- Higher bandwidth and lighter cable. Optical modulation systems have inherently higher data rate capabilities and glass and plastic weighs less than copper. Bandwidth is typically 100 megabits.
- Lower loss than coaxial systems. New low loss fibers extend transmission distance.

- Negligible crosstalk. If each fiber optic channel is optically sheathed there is no crosstalk. Even adjacent unsheathed fibers rarely interfere with each other.
- Ultimately lower cost than either coaxial or twisted-wire systems. The raw material (sand) used in making fiber optics is abundant, while copper gets increasingly more expensive. Cost for a data transmission system is ultimately based on dollars per megabit times distance. Since fiber optic systems have higher bandwidths, the cost factor is slowly moving in their favor.

Key ingredients in any optical communications system are the transmitters and receivers. The ultimate data rate is a function of how fast the transmitter can turn on and off, sending one bit of information, and whether the light sensitive receiver can track this transition. If the data rate is very low, say, 110 bps in your experimental setup, a simple incandescent light and cadmium sulfide photocell will suffice. Higher data rates require much faster response and dictate use of LEDs (light emitting diodes) and phototransistors or photodiodes. Common red LEDs will easily handle 100 K bits per second and most common phototransistors, if properly biased, will also suffice. Higher frequencies require specially fabricated LEDs or, if the transmission line is especially long, then laser diodes might be in order.

It is important to know what each of the components in the system is and the way its selection affects the other components. The designs illustrated in this article are included to demonstrate a workable low frequency system which the personal computer enthusiast may wish to build. The physical elec-

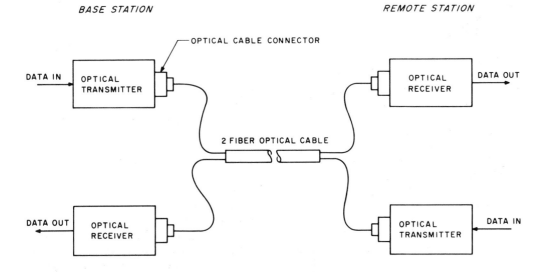

Figure 1: Block diagram of full duplex optical communications link.

tronics of high frequency commercial systems differ considerably, but the physical laws and general concepts are the same.

Fiber Optics

Fiber optics are just what they sound like — glass fibers which conduct light rather than electricity. To understand optical fibers we must look at a few definitions. An example of reflection and refraction is illustrated in figure 2. When a light ray strikes a boundary, partial reflection and partial transmission take place. The materials on either side of the boundary have particular constants n_1 and n_2 respectively (called *indices of refraction*) associated with them. These constants are dependent upon wavelength of the light transmission and the speed of light through the material. Reflection and refraction are related as follows:

$$\text{Reflection } \theta_1 = \theta_1{}'$$
$$\text{Refraction } n_1 \sin \theta_1 = n_2 \sin \theta_2$$

The fiber has a *core*, a light transmitting material of higher index of refraction surrounded by a *cladding* or optical insulating material of a lower index of refraction. Figure 3a is a pictorial representation of a single fiber. Light enters the fiber at an infinite number of angles but only those rays entering the fiber at an angle less than the *critical acceptance angle* are transmitted. Light is propagated within the core of a multimode fiber at specific angles of internal reflection. When a propagating ray strikes the core/cladding interface, it is reflected and zigzags down the core. This is further illustrated in figure 3b.

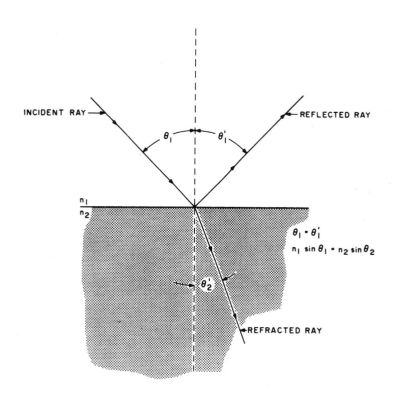

Figure 2: An example of reflection and refraction at an interface, such as the side of the optical cable.

Figure 3: Pictorial diagram of a single fiber illustrating the cladding and core boundary. Only light entering within the "acceptance cone" will be guided down the optical fiber as in figure 3b. Any rays outside this cone are not transmitted.

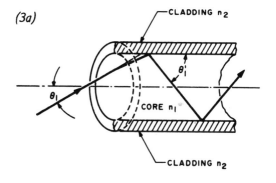

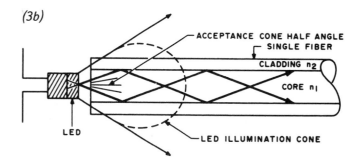

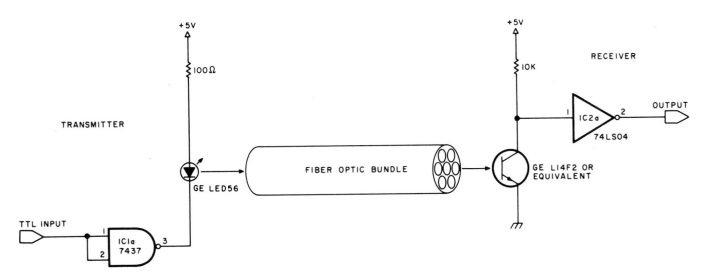

Figure 4: Schematic of a simple low speed and short distance fiber optics communications interface.

Photo 1 demonstrates that a very bright light can be transmitted through a single fiber. In this example the conductor is a single 40 mil plastic fiber with a helium-neon laser as an illumination source.

A fiber optic transmission system using readily available components can be constructed by any interested experimenter. A simple interface is shown in figure 4. An LED driven by a 7437 NAND buffer is focused into the end of a fiber optic bundle. The light emitted at the other end is focused on a phototransistor. When the light strikes the phototransistor it effectively grounds the input of the 74LS04, producing a high output. The connection between the LED, fiber optics, and phototransistor is facilitated through use of special optical connectors. Photo 2 shows an assortment of the type which should be used to build the interface in figure 4.

Lasers

The circuit of figure 4 is useful for only a short distance. This is due primarily to the low intensity of a standard LED. For greater distances a more intense light source is needed. This calls for a device such as a laser, an acronym that stands for *light amplification*

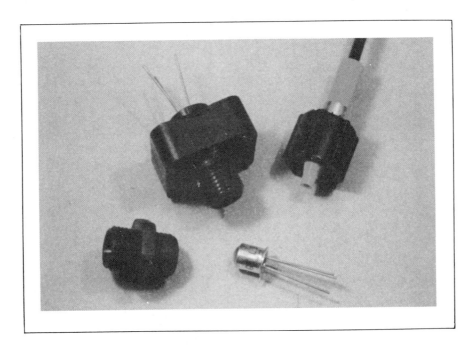

Photo 2: Special connectors necessary to use fiber optics properly. Shown here (starting in the upper right corner and continuing clockwise) are a fiber optic cable with an end connector, a phototransistor in a TO-18 package, an extension coupling which allows two cables to be connected, and a bulkhead receptacle containing either an LED (light emitting diode) or phototransistor.

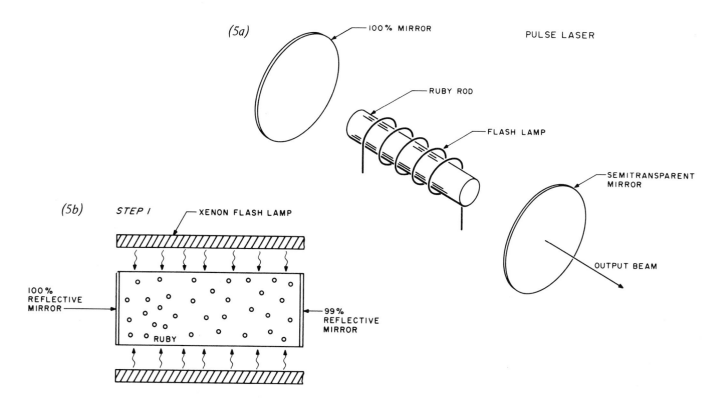

(5a)

100% MIRROR

RUBY ROD

FLASH LAMP

SEMITRANSPARENT MIRROR

OUTPUT BEAM

(5b)   STEP I   XENON FLASH LAMP

100% REFLECTIVE MIRROR

99% REFLECTIVE MIRROR

RUBY

by *stimulated emission* of *radiation*. Light from a laser is all the same frequency, unlike the output of an incandescent bulb. Laser light is referred to as *coherent*, and has a high energy density. It can travel great distances without diverging from a tight beam.

The basic requirements for the creation of a laser are quite simple. We need a material that can absorb and release energy. Next, we need an energy source for exciting this material and a container to hold and control the lasing action, such as a glass tube or solid crystal.

In the actual lasing process, the laser material is placed inside the container, and then stimulated by means of an energy source into the emission of light waves. The laser beam is created by channelling the energy of these light waves into a particular and controlled direction. The result is a highly concentrated, brilliant beam of tremendous power. Figure 5 is a schematic of the first laser invented by Dr Theodore Maiman and a pictorial description of the lasing process.

The ruby laser is a pulse type laser which only produces a light output when the xenon lamp flashes. The best flash lamp can be fired only a few hundred times a second without extensive cooling apparatus. In a ruby laser this pulse mode operation is suitable for cutting stone and welding steel, but not for data communications, because the duty cycle is too short and the energy density too high for low cost fiber optics. The solution is to use a laser that operates continuously, such as a helium-neon gas laser

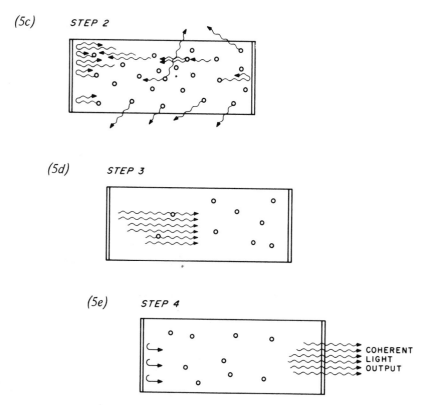

(5c)   STEP 2

(5d)   STEP 3

(5e)   STEP 4

COHERENT LIGHT OUTPUT

Figure 5: The first laser, invented by Dr Theodore Maiman, was made from a ruby rod excited by a xenon flash lamp. A schematic representation is shown in figure 5a. The laser builds up energy by the following process. In figure 5b the flash lamp is fired thereby exciting the electrons in the ruby rod. As the electrons drop back to their original energy level (step 2, figure 5c) they emit photons in random directions. In-step collisions of photons with other excited electrons start a wave front between mirrors (figure 5d). After many reflections back and forth between the mirrors, a wave front is built up until it contains sufficient energy to pass through the slightly less reflective of the two mirrors. This light output consists of coherent light.

SOLID STATE LASER

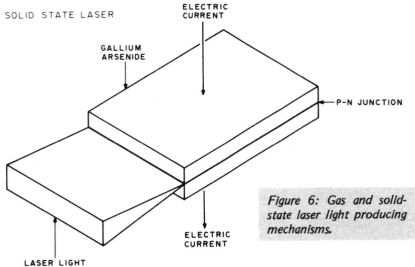

GALLIUM ARSENIDE

ELECTRIC CURRENT

P-N JUNCTION

ELECTRIC CURRENT

LASER LIGHT

(figure 6) or a laser diode which can be pulsed often enough to carry useful data.

The He-Ne laser uses mirrors and electrical excitation in a manner similar to the solid crystal type except that the lasing action is continuous. Photo 3 shows a He-Ne laser in operation. The particular unit has a power output of 2.2 mW and is made by Metrologic Inc. This type of laser can be modulated (the power supply high voltage is modulated) and used to drive a fiber optic bundle, but it is not normally used in that application. The light output of a He-Ne laser is usually red.

Figure 6: Gas and solid-state laser light producing mechanisms.

GAS LASER

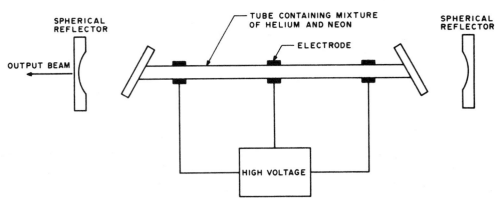

SPHERICAL REFLECTOR

TUBE CONTAINING MIXTURE OF HELIUM AND NEON

ELECTRODE

SPHERICAL REFLECTOR

OUTPUT BEAM

HIGH VOLTAGE

Photo 3: A laser on a tripod shooting across my living room. The laser is a 2.2 mW unit built by Metrologic Instruments of Bellmawr NJ 08031 (this particular model is the ML-969). This picture was taken at night; the trees outside are illuminated by outside flood lamps.

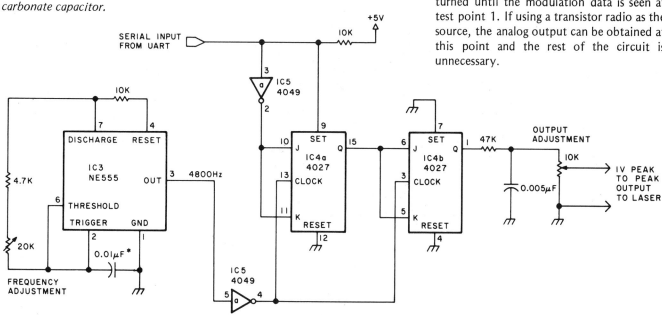

Figure 7: System configuration necessary for one computer to transmit data to another via a helium-neon laser beam. The schematic for the modulator and demodulator are shown in figures 8 and 9, respectively.

Figure 8: A frequency shift keyed laser modulation interface. This circuit accepts input from the computer's UART (universal asynchronous receiver and transmitter). A logic 1 input produces a 2400 Hz output. An input of logic 0 produces a 1200 Hz output. The power connections for the integrated circuits are shown in table 1. The starred capacitor is either a mylar or polycarbonate capacitor.

The most economical high intensity light source for long runs of fiber optics is the laser diode. Don't be so whimsical as to run out and buy one thinking you are going to make a ray gun — it should be just as easy to use as an LED. Laser diodes get very hot in operation and are generally operated only in pulse mode. An 8 W laser diode sold through the surplus dealer can have an average power of only a few hundred microwatts when used in pulse mode operation. Using laser diodes in continuous operation is beyond the talents and resources of most hobbyists and must be left to the commercial ranks for the moment. The light output from a laser diode is infrared and invisible to the human eye.

Communicating on a Laser Beam

While it is possible to demonstrate communication with a laser diode, it is much more dramatic with a He-Ne laser since you can see the beam. A He-Ne laser can be modulated, but it cannot be turned on and off rapidly like an LED or diode. Instead the light intensity is modulated by the data signal. The Metrologic laser I used is a type ML-969 "modulatable" laser. It has a BNC connector on the rear and accepts a 0 thru 1 V input for 0 to 15 per cent intensity modulation. Any greater degree of modulation shuts off the lasing action.

Figure 7 illustrates the system configuration necessary to transmit data from one computer to another. Figure 8 is the schematic of a FSK (frequency shift keyed) modulation interface which can be used as the input to the laser. A 4800 Hz frequency reference produced by IC1 is divided by IC2 to give either 2400 Hz or 1200 Hz for a 1 or 0 logic input respectively. The modulation input to the laser can be any 1 V input up to 500 kHz bandwidth. A transistor radio is a good test source for experiments.

The receiver is shown in figure 9. The laser beam is directed at the phototransistor. With no modulation, the sensitivity is adjusted to set the phototransistor in the middle of its linear range. With the modulation turned on, the trigger adjust control is turned until the modulation data is seen at test point 1. If using a transistor radio as the source, the analog output can be obtained at this point and the rest of the circuit is unnecessary.

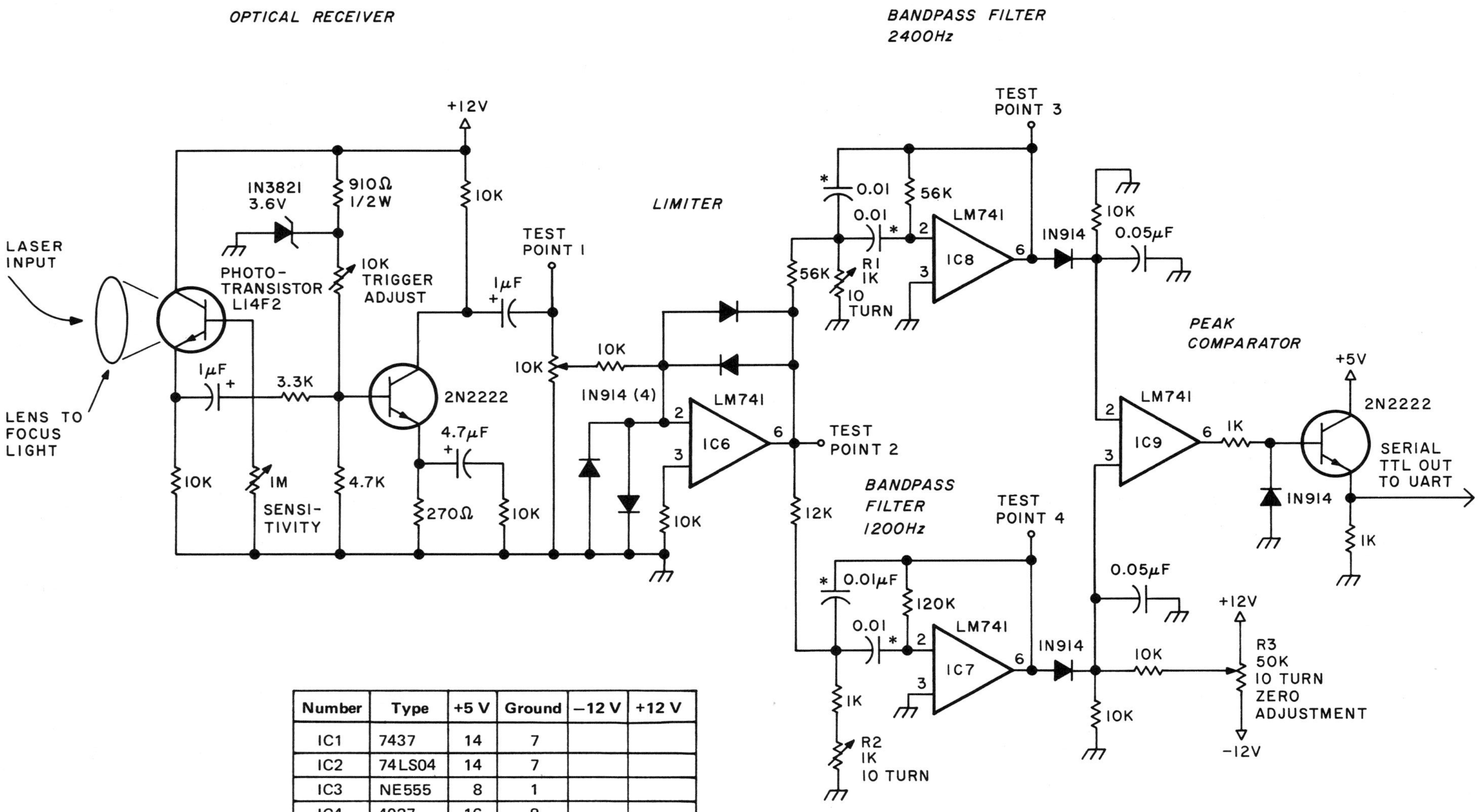

Figure 9: Modulated laser beam serial data receiver. The demodulator consists of two bandpass filters, one for 2400 Hz and the other for 1200 Hz. The power connections are given in table 1. The starred capacitors are mylar or polycarbonate capacitors. All resistors are 1/4 W unless otherwise specified. All diodes are type 1N914.

Table 1: Power pin connections for the integrated circuits used in constructing the laser communicator.

Number	Type	+5 V	Ground	−12 V	+12 V
IC1	7437	14	7		
IC2	74LS04	14	7		
IC3	NE555	8	1		
IC4	4027	16	8		
IC5	4049	1	8		
IC6	LM741			4	7
IC7	LM741			4	7
IC8	LM741			4	7
IC9	LM741			4	7

Figure 10: A triple voltage power supply for the laser modulator.

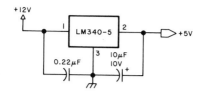

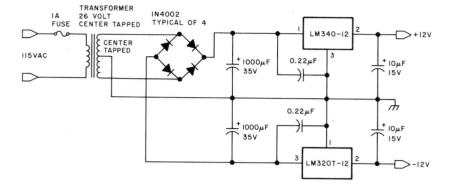

Integrated circuits 1 thru 4 form a frequency shift keyed demodulator with a TTL (transistor-transistor logic) output which is sent to a UART (universal asynchronous receiver-transmitter). To tune this section, first connect a 1200 Hz signal source to test point 1. Turn potentiometer R2 until the output amplitude of IC3 test point 4 peaks. Then apply 2400 Hz to test point 1 and adjust R1 until the amplitude at test point 3 also peaks. R3 adjusts the point at which circuit's output switches between logic levels. It should be set to follow the input at test point 1 with the shortest response time.

While the 15 per cent modulation could be detected directly and converted to NRZ (nonreturn to zero) formatted data, the receiver circuitry would be far more complicated. The combination of amplitude and frequency modulation techniques is intended to add significantly to the chances that an experimenter will have success building it. The critical parameters (as with any optical system) are alignment and light level. And, while you may never have to transmit a Red Sox baseball game across two miles of New Hampshire woods, it's nice to know how if you ever have to do it.∎

LONG DISTANCE COMMUNICATION

Dear Steve,

I saw your article "Communicate on a Light Beam," and became very interested. I have an application which requires sending data up to a kilometer at speeds from 2000 to 9600 characters per second (cps). Your descriptions of the fiber optic cable and the light-emitting diode (LED) transmission circuits seem to be ideal, if they are cost-effective.

Could you give more details of the distances which the circuits can drive and the addresses of the suppliers of the fiber optic components?

R H Fields

Realize, of course, that the circuits presented, while possibly usable in commercial applications, are presented more to introduce the reader to the concept of fiber optic communications than solve any particular application problem. Their usability in a 1 kilometer data link depends upon more than just the electronic parameters of the circuit. The laser probably can drive such a length, but cable losses and mechanical/optical connections are going to be an important factor in any success.

When you speak of 9600 cps, that is approximately 100 k bits per second (bps) and is a reasonable transmission rate. However, response time of the receiver electronics is going to be much more critical than a 10 k rate. Given the length of cable as 1 kilometer, I would caution you that a certain in-tensity must be maintained at the output to achieve this response.

Rather than try to reinvent the wheel or try to second-guess the technical people who really know the field, I think you would be better off purchasing a commercial system. The following is a list of American companies that deal in fiber optics. I am sure they will have a cost-effective solution for you:

Corning Glass Works
Telecommunications Dept
Corning, NY 14830
(607) 974-8812

Dupont Co
Plastic Products and Resins Dept
Wilmington, DE 19898
(302) 774-7850

Fiberoptic Cable Corp
POB 1492
Framingham, MA 01701
(617) 875-5530

Galileo Electro-Optics Corp
Galileo Park
Sturbridge, MA 01618
(617) 347-9191

General Cable Corp
500 W Putnam Ave
Greenwich, CT 06830
(203) 661-0100

ITT
Electro-Optical Products Div
Roanoke, VA 24019
(703) 563-0371

Quartz Products Corp
688 Somerset St
Plainfield, NJ 07061
(201) 757-4545

Times Fiber Communications Inc
358 Hall Ave
Wallingford, CT 06492
(203) 265-2361

Valtec Corp
Electro Fiberoptics Div
West Boylston, MA 01583
(617) 835-6083

For further descriptive information on the use of fiber optics, I suggest you refer to the January 5, 1978 issue of EDN magazine and an article entitled "Designer's Guide to Fiber Optics."...Steve

Ciarcia's Circuit Cellar

Mind Over Matter

Add Biofeedback Input to Your Computer

I wouldn't want you to get the wrong idea from photo 1. I haven't given up computers and taken up telling fortunes. Just consider the photo as a slightly dramatized introduction to a topic we've all heard of, but know so little about: biofeedback. In layman's terms, this simply means having the capability to monitor (in this case electronically) physiological processes.

There are a variety of devices on the market referred to as brain wave monitors. Brain waves are but one of the many sources of energy categorized under biofeedback. Their common relationship is that they are all electrical pulses which run through the body as a result of brain or muscle activity. Nerves and muscles within the body generate

electricity by electrochemical action similar to that in a battery.

When we want to lift an arm, the brain sends an electrical pulse to the muscles in the arm. Proper magnitude and duration of the signal result in coordinated activity. The actual energy that is transmitted from the brain is very small: on the order of a few hundred microvolts at the most. The most familiar of these signals is the voltage generated by the pumping of the heart. A graph of this voltage versus time is called an electrocardiogram (abbreviated EKG or ECG). An EKG looks like a spiked waveform, with periodic response equivalent to a heartbeat. Many individual muscle contractions contribute to a frequency spectrum of 0.1 to 100 Hz, with an amplitude of about 5 mV.

Another group of signals are the voltages generated using large skeletal muscles like biceps and triceps. A recording of these voltages is called an electromyograph or EMG. Occurring only when the muscles contract, not periodically like the heart, the frequencies are very low, but the voltage is higher: about 5 to 10 mV. Because of their magnitude, these signals are the easiest to monitor.

The last important biomedical signal is composed of very low amplitude voltages within the brain itself. These are recorded by the EEG (electroencephalograph). They exhibit both periodic and pulse mode. The 50 μV signals occupy a band that is generally between 1 and 30 Hz. The signals are further subdivided into delta, theta, alpha, and beta waves. These classifications signify activity in defined frequency bands. Differences in activity seem to reflect particular personality tendencies.

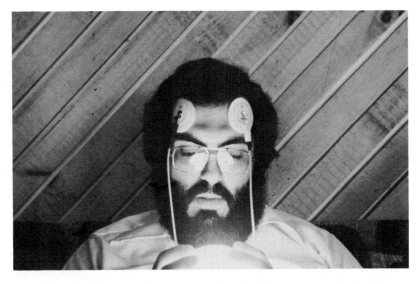

Photo 1: This photo simulates a crystal ball reflection to emphasize the control capabilities associated with this article.

Monitoring Internal Electrical Activity

Consider the activity within the brain or the muscles. Each neuron is producing minute voltages. In combination with the voltages of billions of other cells involved in similar activities, the result is fairly significant. The situation can be compared to that of a football stadium before, during, and after a game. A listener outside of the stadium would not hear the shouts of a few individuals, but 50,000 people shouting is quite another story. A further consideration is the progress of the game. Loud noise coming from a particular section of the stadium during the game signifies approval. This same ovation at the conclusion can imply the identity of the winner. Observation and association are the keys. EKG, EMG, and EEG readings must be carefully interpreted.

All of the signals discussed thus far can be monitored with surface electrodes. When the biceps is moved, a small voltage which can be measured will be produced across it (ie: referenced to some other point on the body). Monitoring this voltage requires a special amplifier with extremely high input impedance and 60 Hz rejection. Care must be taken to use a device which will not load the signal being sensed, nor have such a low signal to noise ratio that one cannot discern intelligible information. The unique device which satisfies these requirements is called an instrumentation amplifier. Any product which is sold to monitor brain waves, EKGs, etc will contain an instrumentation amplifier.

Instrumentation amplifiers are often called differential or data amplifiers. They are closed loop gain blocks with accurately

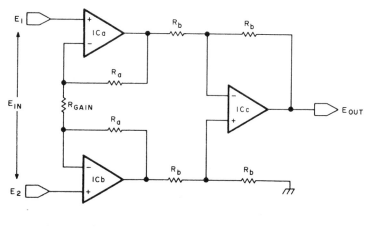

$$E_{OUT} = \left(1 + \frac{2R_a}{R_{GAIN}}\right)\left(E_2 - E_1\right)$$

Figure 1: Differential input instrumentation amplifier configured from multiple, single-ended, operational, amplifier elements.

predictable input to output response. They are especially configured to have extremely high input impedances and common mode rejection which makes them ideal for amplifying low level signals in the presence of large common mode voltages. Figure 1 shows the schematic of a typical instrumentation amplifier built from such standard operational amplifiers as LM301s or 741s.

This common circuit consists of three op amps. ICa and ICb are inserted as high impedance input buffers which provide a differential gain of $1 + 2R1/R_{gain}$ and unity common mode gain. ICc is a unity gain differential amplifier which combines the voltages from the other amps. The ratio of the differential voltage gain of an amplifier to its common mode gain is enhanced

Table 1: Comparison chart of three different amplification elements.

	Operational Amplifier	Instrumentation Amplifier	Isolation Amplifier
Symbol			
Feedback Configuration	1. User defined feedback such as voltage or current. 2. Can be configured to provide dV/dt, ∫Vdt, log V, etc.	1. Committed feedback. 2. Gain adjustable within fixed limits.	1. Committed feedback. 2. Gain adjustable within fixed limits.
Basic Applications	1. General purpose amplification element. 2. Buffer. 3. Analog computational element.	1. High accuracy analog sense amplifier.	1. High accuracy analog sense amplifier. 2. Analog safety isolator. 3. Prevents ground loops.

by selecting low feedback resistors to reduce the effects of input offsets. A problem arises when selecting matched components to build this otherwise cheap circuit. Slight variations in resistors and op amps can make the difference between a working or nonworking circuit. (More on that subject will be discussed later.)

EEG and EMG monitoring requires an instrumentation amplifer because of the low input levels; but, when used in a biomedical application, a further modification to the amplifier's internal design is necessary. The special device is called an isolation amplifier. Transformers or optical couplers inside the amplifier block isolate the sense inputs of the amplifier from the output circuitry. This means that a 2 μV signal could be monitored on a 2000 V transmission line and the output connected directly to an analog to digital converter input on your computer. The protection works both ways. This is why any connections to the body are done through isolation amplifiers.

An isolation amplifier is to analog signals as an optoisolator is to digital signals. It prevents ground loops from the data analysis equipment (ie: your computer) through the subject. When the electrodes are attached, skin contact resistance is very low: only a few hundred ohms. A leakage current of just 100 μA can be fatal. Table 1 summarizes the differences between the amplifiers we've discussed.

Choosing an Isolation Amplifier

There have been many articles on the subject of alpha brain wave and muscle monitors; some even include circuit diagrams for construction of the interfaces. The major thing these articles lack is a caution about matching components, and the critical importance of proper layout. The circuit of figure 1, if breadboarded in the usual fashion, wouldn't have a chance of working on 50 μV levels. Even the testing of a handful of components to obtain matched pairs would be useless without concise wiring and plenty of ground plane shielding to reduce 60 Hz interference. Personally, I don't like to present circuits with so many strings attached that it takes divine intervention to make them work.

The final most important consideration in this undertaking is to not get electrocuted because of sloppy technique. At this point I'd like to draw the line between this article and other construction oriented articles. A cheap method of attaining minimal isolation is to use batteries to power an instrumentation amplifier. This sounds fine in theory, but it is very risky in practice. Too often a standard power supply is substituted for the batteries, or a loosely wired component falls against a live wire on another circuit.

Fortunately we can get both safety and performance if we don't assume that everything has to be constructed from scratch. It is a much better idea to take advantage of commercially available isolation amplifiers. (You wouldn't build a 4 bit digital counter from transistors, would you?) A perfect choice for this application is the Underwriters Laboratory approved Analog Devices 284J isolation amplifier shown in photo 2. It provides plus or minus 2500 V isolation, 110 dB common mode rejection, and a gain of 10 V per volt. For the experimenter this eliminates building the only tricky section of the interface. An added benefit is that the isolation is now an internal function of the 284J and not a function of installation. The ultimate aim of this article is to produce a biofeedback interface for a computer; I don't want anyone getting injured in the process.

Biofeedback Computer Interface

Figure 2 is the schematic of a circuit which is capable of sensing the minute voltages we've been discussing, and signifying to the computer when a present level has been attained. This is a bare bones, basic interface designed specifically for signal acquisition. It would seem to me that this is the area which would give most people problems. The circuit consists of an isolation amplifier module, two gain stages, and a comparator to sense peak level. The completed circuit is shown in photo 3.

All connections to the body are done through M1. The high and low input ter-

Photo 2: The Analog Devices 284J isolation amplifier used in this article.

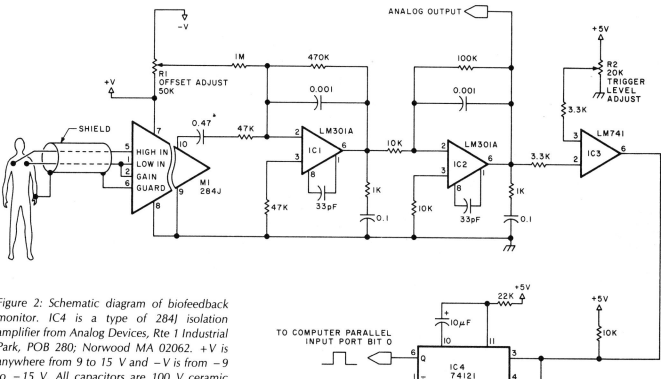

Figure 2: Schematic diagram of biofeedback monitor. IC4 is a type of 284J isolation amplifier from Analog Devices, Rte 1 Industrial Park, POB 280; Norwood MA 02062. +V is anywhere from 9 to 15 V and −V is from −9 to −15 V. All capacitors are 100 V ceramic unless otherwise noted. All circuitry should be mounted on a ground plane to reduce AC pickup. Connecting wires should be as short as possible. The electrode cable must be shielded to obtain proper operation.

Number	Type	+5 V	GND	+V	−V
IC1	LM301A	−	−	7	4
IC2	LM301A	−	−	7	4
IC3	LM741	7	4	−	−
IC4	74121	14	7	−	−

Table 2: Power pin connections for figure 2 schematic.

minals are attached across the area to be monitored. If it is an EKG output, you should attach the terminals as shown. For biceps input, these two probes would go on the upper arm and the guard connected to the wrist. All leads between the body and the board must be shielded or 60 Hz will be all that is seen on the output. Gain on the 284J amplifier is set by connecting a resistor between pins 1 and 2. When they are shorted as shown, the result is a gain of 10.

ICs 1 and 2 are configured as common inverting amplifiers, each having a gain of 10. Since the signals we want to amplify are relatively low frequency AC, a capacitor is attached at the input of the first amplifier to filter out the DC component of M1's output. In most cases of muscle monitoring, this total gain of 1000 is sufficient. Picking up brain waves will require additional amplification. Changing the 100 kΩ resistor on IC2 to 1 MΩ will increase it another order of magnitude to 10,000. Be aware that raising the amplification also raises the noise on the output. Capacitors in the feedback loops are used in an attempt to keep this noise to a minimum. The amplified analog signal is available at pin 6 of IC2.

Photo 3: View of the prototype circuit described in figure 2.

Photo 4: Pregelled American Optical electrodes of the type used in this article. They are available from medical supply outlets.

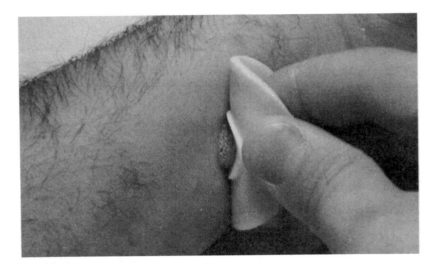

Photo 5: The electrode has a saturated spongy center which serves to reduce skin contact resistance. It is necessary to use this type of connection to the body if satisfactory results are to be obtained.

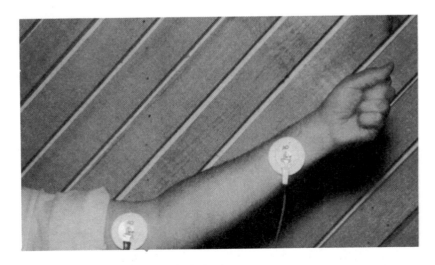

Photo 6: To monitor the electrical activity of the muscles in the arm, electrodes should be placed as shown.

It can be attached to an oscilloscope if you care to watch yourself in action.

IC3 and IC4 are the interface to the computer. IC3 is a comparator with normally high output. When the signal level from IC2 exceeds the trigger voltage set on R2, IC3 pin 6 goes low, firing the one shot IC4. This signal is in turn connected to a parallel input bit of the computer. Offset potentiometer R1 is adjusted to give 0 V on IC2 pin 6 when M1 is removed and M1 pin 10 is grounded.

Using the Muscle Monitor

Monitoring muscle voltages is much easier than monitoring brain waves. To adequately accomplish the latter, sharp band-pass filters which can separate brain waves from other signal sources must be added to figure 1. As it stands, it cannot differentiate between alpha or theta waves but is optimized for muscle pickup.

To sense the electrical activity of a muscle such as the biceps, three electrodes are necessary. It is not enough to merely wrap three wires around your arm. Special electrodes such as the type shown in photo 4 are necessary. These are referred to as pregelled silver-silver chloride disposable electrodes and they are available through medical supply outlets. The electrodes (shown in photo 5) have a spongy center section saturated with a gel to reduce skin contact resistance. The best results will be obtained by using these or similar attachments.

In the case of the forearm muscles, the high electrode (shown in photo 6) is placed on the wrist, the low electrode on the upper arm, and the guard on my chest, close to the shoulder. When the muscles of that arm are flexed, a large pulse will appear at the analog output terminal of the interface. It is best seen with an oscilloscope. Every movement produces some noticeable deviation in the trace. If the trigger adjustment R2 is set above the ambient noise at the peak of this large pulse, it will fire the one shot every time the muscle is flexed. Actually, adjustment can be much finer. With the electrodes placed as in figure 1 (the guard is on my chest again), they can pick up something as insignificant as moving your eyebrows or gritting your teeth. The setting is made higher than the level produced when talking or breathing, so that it can be used as a suitable control input to the computer.

Biofeedback Computer Control

Control is the name of the game. Consider someone who is almost totally paralyzed. This system could be used (perhaps by sensing eyebrow movement) as an on/off switch to a more sophisticated controller. I've seen one computer aid for the handicapped which consisted of an alphanumeric sequencing display. Letters could be individually chosen and eventually combined to produce whole written messages. A lot can be accomplished with a single bit of input if the software is written with time as a pertinent consideration. A single switch could signify a particular choice if each was presented in sequence with time allotted to answer. That is the premise of the BASIC program in listing 1.

This is a simple program written in Micro Com 8 K Zapple BASIC. It presents the operator with a series of seven choices, and branches to special subroutines as a result of these choices. It presumes that the user can see and signify positive response by a high logic level on bit 0 of input port 3. This bit is tied to the output of our eyebrow twitch monitor. Output port 17 has seven lights attached to bits 1 through 7 (bit 0 not used). The program lights the first light, and the user decides whether or not the computer should perform the activity signified by bit 1. If so, the user merely furrows his or her brow and the program jumps to the designated activity. In this simple illustration, I merely flash the light a few times to indicate which was chosen. Should the operator not care for the first choice, the program sequences to the next choice, and so on. Before hookup, the program can be easily tested with the muscle monitor by temporarily attaching a normally closed, pushbutton switch on port 3 bit 0.

Conclusion

All of this effort for a single bit of data acquisition may appear unjustified, but it can prove to be exceedingly significant in situations where no other means of computer interaction is available. At the least, the interface should provide a substantial base for biofeedback experiments. With additional amplification and filtering to monitor brain waves, a whole series of challenging experiments come to mind. Personal computing need not be relegated to the level of canned amusements and commercial presentations. A refinement of this interface could be the one critical design feature which would open the field of personal computing to individuals who are otherwise physically unable to take advantage of it.■

```
100 REM This program demonstrates how the computer can be
110 REM used to provide contol output from an EMG digital input
120 REM EMG input is on port 3, bit 0.  No stimulus is logic 0
130 REM while muscle activity is signified by logic 1.
140 REM Test apparatus uses 7 lights attached to bits 1 thru 7 of
150 REM output port 17. The computer sequences thru the lights until the
160 REM operator signifies a choice by --"THINKING"-- about it !!!
170 REM
180 REM Copyright 1979    STEVE CIARCIA
190 REM
200 REM
210 FOR D=0 TO 300 :NEXT D
220 REM
230 REM This routine sequentially flashes bits  1 through 7 of port 17
240 REM It only exits when an input flag has been set by the EMG monitor
250 B=1
260 X=2^B :OUT 17,X
270 GOSUB 440
280 IF F=1 THEN OUT 17,1 :GOTO 320
290 B=B+1 :IF B>7 THEN GOTO 210
300 GOTO 260
310 REM
320 IF B=1 THEN GOSUB 670 :GOTO 570
330 IF B=2 THEN GOSUB 670 :GOTO 580
340 IF B=3 THEN GOSUB 670 :GOTO 590
350 IF B=4 THEN GOSUB 670 :GOTO 600
360 IF B=5 THEN GOSUB 670 :GOTO 610
370 IF B=6 THEN GOSUB 670 :GOTO 620
380 IF B=7 THEN GOSUB 670 :GOTO 630
390 IF B>7 THEN STOP
400 REM
410 REM
420 REM This routine reads the EMG monitor on port 3 bit 0
430 REM If signal is present it sets flag F=1
440 A=0 :F=0
450 I=INP(3)-254
460 IF I>0 THEN 490
470 A=A+1 :IF A>200 THEN RETURN :REM give operator time to respond
480 GOTO 450
490 F=1
500 Q=INP(3)
510 IF Q>254 THEN 500
520 RETURN
530 REM
540 REM
550 REM These 7 routines can be replaced with outputs to
560 REM individual control programs.
570 PRINT"b=1":GOTO 210
580 PRINT"b=2":GOTO 210
590 PRINT"b=3":GOTO 210
600 PRINT"b=4":GOTO 210
610 PRINT"b=5":GOTO 210
620 PRINT"b=6":GOTO 210
630 PRINT"b=7":GOTO 210
640 REM
650 REM
660 REM This routine flashes individual light to indicate selection
670 FOR T=0 TO 10
680 OUT 17,X
690 FOR T1=0 TO 50 :NEXT T1
700 OUT 17,0 :
710 FOR T1=0 TO 50 :NEXT T1
720 NEXT T
730 RETURN
```

Listing 1: BASIC program to sense input from the biofeedback monitor. This program scans the cursor through several choices and waits a short period of time. If the user squints or blinks within the allotted period, that choice is designated. If it is not designated, it cycles to the next choice. This particular program just blinks the chosen objective to indicate that the interface is working. The required body connections for picking up eyebrow movement are shown in photo 1.

MIND OVER MATTER EXPANSION

Dear Steve,

I found your article "Mind Over Matter" very interesting. When all the components arrive, I hope to have an operational muscle monitor. A friend of mine has a great deal of enthusiasm for brain wave monitors, and, although I do not quite see the magic he sees in them, the idea is intriguing.

My difficulty with building the brain wave monitor is that my knowledge of electronics has never got past reading the Heathkit instructions stage. You mentioned changing the 100 K ohm resistor on IC2 to 1 M ohm for brain wave amplification, which is OK; however, you then said that band-pass filters must be added, and you have lost me.

I know it would be a time-consuming project, but I thought that I would try and trouble you for a circuit and parts list at the Heathkit-level for brain wave monitor expansion. I assume that, along with input to an oscilloscope (Heathkit, naturally), the analog output could be used as input to my Cromemco D + 7A I/O board?

Frank Gizinski

I hope you will have an operational muscle monitor by the time you read this. I regret, however, that I cannot comply with your request. Heathkit and the Muppets both have something in common: because the original is done so well and anything equivalent could only be accomplished by a similar effort, there are no copies. Except through the effort of a complete article on the subject, I hesitate to do only half the job by sketching out a few filter circuits which ultimately demand a great deal of technical ability.

In addition to yours, many letters have requested expansion information. In actuality, the required circuitry would constitute a low-frequency spectrum analyzer. I will look into the design, and use it either as an article specifically on expansion of the "Mind over Mat-

ter" introduction, or as an additional supplement with one of my regular monthly offerings. I am aware of the obvious interest in expansion, and I do try to present circuits that can be readily constructed.

Finally, the biofeedback interface can be readily used with the Cromemco A/D board, if the analog output from the monitor is scaled down to 0 to 2.56 V. This can be done with a 500 K ohm potentiometer serving essentially as a volume control. Analysis of the acquired data is another subject entirely.

*Perhaps your strength is really software, and you will achieve success better by this method. The ultimate goal is to analyze the low-frequency spectrum. This can be done either through hardware or software....***Steve**

OPERATIONAL AMPLIFIERS

I have been using the AD284J isolation operational- amplifier system that you described in "Mind Over Matter" as an EKG (electrocardiogram) monitor, in conjunction with a surplus chart recorder. Can you recommend some books that will help me to learn more about operational amplifiers?

**Matsutoshi Uchiyama
Tokyo, Japan**

I am glad you are gaining experience with the circuit. As far as expanding your mind a little, I suggest the following books:

- Operational Amplifiers —Design and Applications, Jerald G Graeme, Gene E Tobey, and Lawrence P Huelsman, McGraw-Hill Book Company, New York NY 1971.
- Applications of Operational Amplifiers—Third Generation Techniques, Jerald G Graeme, McGraw-Hill Book Company, New York NY 1973.
- Handbook of Operational

Amplifier Circuit Design, *David F Stout and Milton Kaufman, McGraw-Hill Book Company, New York NY, 1976.*

*I hope these help....***Steve**

EMG + TRS-80 = ??

Dear Steve,

I am currently using a TRS-80 Level II 16 K microcomputer in my classroom. I am a Special Education specialist who teaches 7th and 8th grade learning-disabled students. I am trying to put together a program using stress-free learning techniques. What I would like to do is interface an EMG (electromyogram) unit to the TRS-80. Your name was given to me as a possible resource. I would appreciate any assistance that you could provide.

William Engelhardt

It is not particularly difficult to connect the single-bit output of the EMG unit from my article "Mind Over Matter: Add Biofeedback Input to Your Computer" to a TRS-80, if you have the Radio Shack Expansion Interface or a COMM-80. Either unit provides a printer port at memory address hexadecimal 37E8.

The easiest method is to attach the EMG output to pin 21 of the printer connector (ground is on pin 34). This is ordinarily used as the printer BUSY line. Pins 23, 25, 28, 29, 19, 32, and 30 should be grounded. In BASIC, execute a PEEK(14312) when you want to read the EMG input. If it returns as decimal 128, then the EMG output is high; if it returns as 0, then its output is low.

*If you would prefer not to go through the expense of the expansion interfaces for a single-bit input, then I refer you to my article "I/O Expansion for the Radio Shack TRS-80, Part 1: Principles of Parallel Ports," which describes how to construct a parallel port for any address....***Steve**

Sound Off

Creating music and sound effects with a microcomputer is an arduous task when the processor must directly synthesize each wave form. The usual technique employed is for the computer to calculate a mathematical model of a desired sound and output it through a digital to analog converter. In theory this is fine, but in practice it requires an extremely fast computer to form complex waves. For example, to synthesize a simple 8 kHz tone, the computer must generate an audio wave coordinate every 62 μs. Use of memory tables to replace some calculations can speed up the process, but the production of complex waveforms or higher frequencies would monopolize all of the processor's available time.

A second technique for sound synthesis is to use an analog approach. The computer can simulate an electronic organ by attaching separate tone generators to the computer which are turned on and off digitally. In *The Toy Store Begins at Home* (page 43) four oscillators were attached and could be individually con-

trolled through an output port. Although a tune with four notes isn't very appealing, it served a purpose, and easily demonstrated this alternative synthesis technique. Complex sounds, such as a musical chord, were created by simply turning three of the tone generators on simultaneously. Unfortunately, the preset frequencies allowed only one chord, and in order to change it the circuit would have to be physically altered.

The concept of the external oscillator is the important fact to point out because the production of the sound no longer presents critical real-time operation to the computer. To further simplify this approach and reduce the necessity for N number of oscillators to produce N conceivable tones, we can design this external generator to be frequency programmable by controlling the timebase components. More on this later.

More often than not, the sounds we hear are not pure tones, but rather are complex combinations of frequencies that are sometimes mixed with noise. In many instances it is the characteristic presentation or amplitude variations rather than the frequency content which we recognize as the relevant quantity.

Photo 1 shows a steam engine. For the model railroading buffs out there, it is a Lionel Southern Crescent steam engine. The chug-chug sound we all associate with trains is nothing more than white noise which is modulated. The amplitude, or envelope, is pulsed on and off to produce the characteristic sounds of a steam locomotive. While, in theory, the computer can directly synthesize all of these sounds, the personal computing enthusiast might find it more rewarding to consider a hardware alternative.

Fortunately, Texas Instruments and General Instrument have come to the rescue with LSI (large scale integration) sound generator integrated circuits. These integrated circuits contain the basic elements of sound synthesis: VCO (voltage controlled oscillators), mixers, envelope generators, noise generators, etc. The Texas Instruments unit is specifically designed to be used independently with sound defined through component selection. The General Instrument unit

Photo 1: Sound effects for a Lionel Southern Crescent model steam engine is one use of a programmable sound generator device.

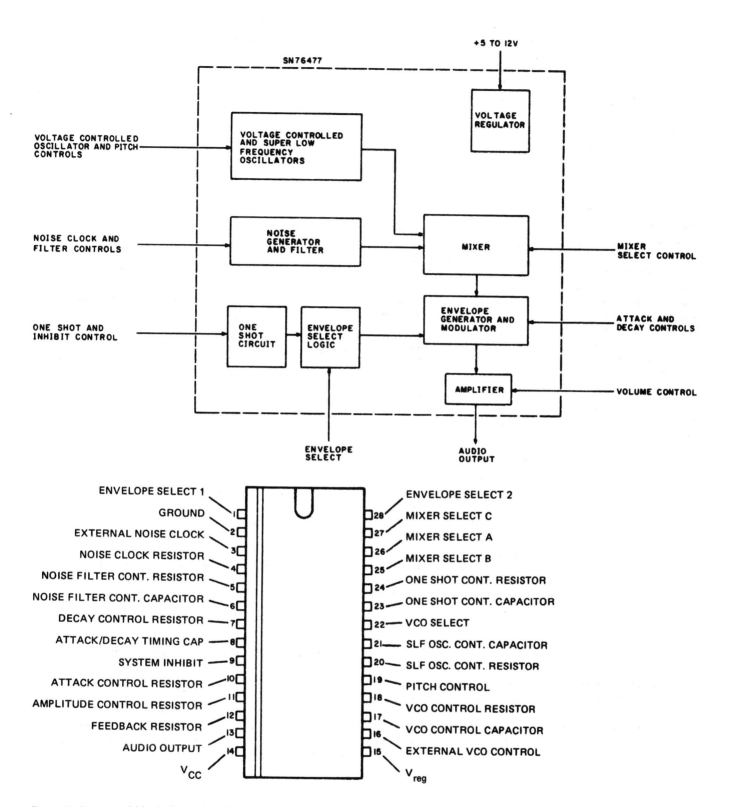

Figure 1: Functional block diagram and pin description of Texas Instruments SN76477 complex sound generator.

is bus oriented and attaches to a microprocessor. Both produce sound, but their interfaces are quite different.

The SN76477 Complex Sound Generator

The SN76477 complex sound generator produces sounds by the value selection of externally attached resistors and capacitors. Internally, as shown in figure 1, the generator contains two voltage controlled oscillators, a noise generator, envelope generator and modulator, and mixers.

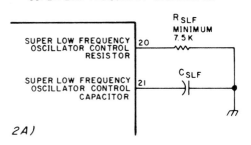

SUPER LOW FREQUENCY OSCILLATOR

2A)

Figure 2: The SLF (super low frequency) and VCO (voltage controlled oscillator) sections of the Texas Instruments SN-76477 complex sound generator. The desired frequency is selected by adjusting the resistor and capacitor circuits. The frequency is determined by the following formulas: super low frequency = $0.64/R_{SLF} \times C_{SLF}$ and voltage controlled oscillator $= 0.64/R_{VCO} \times C_{VCO}$.

Figure 3: The noise generator and filter section is composed of an external clock, two resistors, and a capacitor. The nominal value of R_N is 47 k ohms and the minimum value for R_{NF} is 7.5 k ohms.

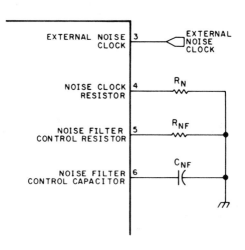

VOLTAGE CONTROLLED OSCILLATOR

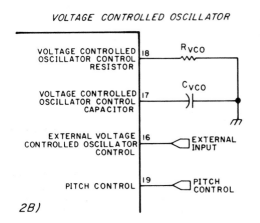

2B)

Oscillators

Figure 2 illustrates the two oscillator sections and equations for frequency selection. Figure 2a is an SLF (super low frequency generator) with a normal range of 0.1 Hz to 30 Hz. This super low frequency output is most often used to provide the input to the voltage controlled oscillator which runs at a higher frequency. Such a combination results in frequency modulated sound synthesis. A familiar example is a siren.

The voltage controlled oscillator can be externally controlled by grounding pin 22. The frequency is then governed by a 0 to 2.35 V signal applied to pin 16. Signals above 2.35 V will saturate oscillator output. As a further enhancement, the voltage controlled oscillator allows pitch control through a similarly ranged signal applied to pin 19.

The output of the voltage controlled oscillator and the super low frequency oscillator is a square wave which is supplied to the mixer and through the envelope selection logic to the envelope generator and modulator.

Noise Generator and Filter

Since so many sounds incorporate noise as an integral component, the 76477 includes a noise generator which can be set to produce pink or white noise by selection of the proper components. (Pink noise has a spectral intensity inversely proportional to frequency over a specified range. White noise is random and has constant energy for a unit bandwidth.) Further refinement of the desired noise range is accommodated through an external clock input applied to pin 3. Figure 3 illustrates this hookup.

The noise generator output is sent to the mixer.

The Mixer and Envelope Selection

Figure 4 shows how the mixer section of a sound generator works and specifically details the logic codes for the SN76477. The mixer is essentially a gating network which digitally combines the outputs from the super low frequency oscillator, voltage

Figure 4: Outputs of the super low frequency oscillator, voltage controlled oscillator, and noise generator are digitally selected. The control table indicates the output produced for any particular input.

Mixer Select Inputs			
C (Pin 27)	B (Pin 25)	A (Pin 26)	Mixer Output
L	L	L	VCO
L	L	H	SLF
L	H	L	NOISE
L	H	H	VCO/NOISE
H	L	L	SLF/NOISE
H	L	H	SLF/VCO/NOISE
H	H	L	SLF/VCO
H	H	H	INHIBIT

H = high level
L = low level or open

controlled oscillator, and noise generator through a 3 bit code applied to pins 25, 26, and 27. An additional inhibit state is added to shut off operation of the mixer when desired.

The individual outputs of the voltage controlled oscillator, super low frequency oscillator and noise generator are selected with codes of 000, 001, and 010 respectively, as shown in the chart accompanying figure 4. The true value of this device is demonstrated when complex sounds are produced by combining these three sources and utilizing the inhibit for emphasis.

Figure 5a shows how the voltage controlled oscillator can be modulated by the super low frequency oscillator. As mentioned, an example of this is a siren. If, on the other hand, the super low frequency oscillator were programmed as in figure 5b, and mixed with the noise generator, the mixer output would sound like the steam engine we previously discussed. For faster on/off pulsing of the noise generator, the voltage controlled oscillator could be selected, and would appear as in figure 5c.

The inhibit line, rather than being an actual sound source, controls the duration of the other three sections. The internal one shot, triggering a 100 ms burst of noise to a loud amplifier, will sound like a gun shot. This is detailed in figure 5d.

The combined mixer output then goes to the envelope generator and modulator where the amplitude (volume) of the output signal is tailored through proper attack and decay timing so that it will synthesize actual sounds accurately. A piano is most easily characterized by its sharp attack and very long decay. Figure 6 outlines the component calculations for these timed functions.

Manual Sound Synthesizer

The SN76477 is essentially an independent sound generator. This means that with a few discrete components it can independently synthesize the sound of sirens, phasers, guns, etc. A computer is not required to program this device and, in fact, with the exception of the envelope, mixer and inhibit selection inputs, it is not directly controllable with a microprocessor. An example of a typical hardwired circuit using the SN76477 is shown in figure 7. This circuit simulates the sound of a steam engine and a whistle. The timing components were selected by using the equations outlined in figures 2 thru 6. This circuit produces two sounds by multiplexing the mixer between the voltage controlled oscillator frequency and the super low noise outputs. Normally, with the push button open the super low frequency oscillator pulses the noise generator on and off, producing a chug-chug sound. When the button is pushed, oscillator IC2 multiplexes the integrated circuit to the voltage controlled oscillator only position approxi-

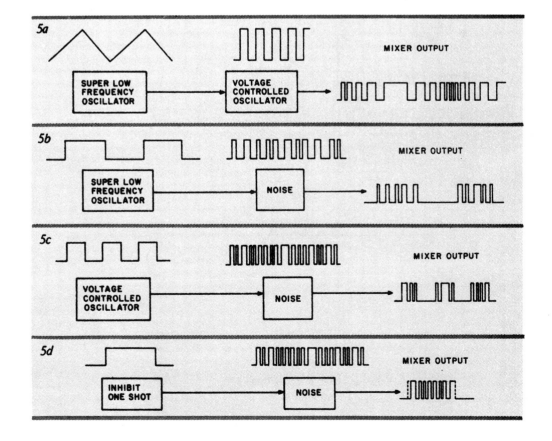

Figure 5: By carefully choosing what signals are combined, a variety of different types of sounds can be produced. Figure 5a shows a combination of the super low frequency generator and the voltage controlled oscillator producing a sound such as a siren. Figure 5b combines the super low frequency oscillator and the noise generator to generate a sound such as a steam engine. In figure 5c, the voltage controlled oscillator and noise generator are mixed together to form a faster on and off pulsing than produced using the super low frequency generator. When the inhibit one shot is mixed with noise (figure 5d) the resulting sound would resemble a gun being fired.

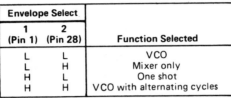

ONE SHOT $\approx$ .8 $(R_{OS})(C_{OS})$

DECAY TIME $\approx (R_D)(C)$

ATTACK TIME $\approx (R_A)(C)$

Figure 6: The envelope selection (table) is determined by envelope select 1 and envelope select 2 (pins 1 and 28) as shown in the table. The attack and decay timing is determined by R_{OS}, C_{OS}, R_D, C, and R_A.

Envelope Select		
1 (Pin 1)	2 (Pin 28)	Function Selected
L	L	VCO
L	H	Mixer only
H	L	One shot
H	H	VCO with alternating cycles

H = high level
L = low level or open

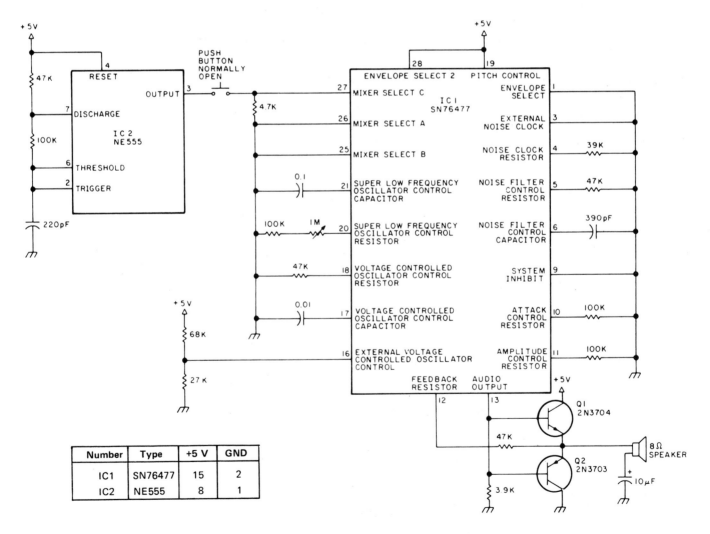

Number	Type	+5 V	GND
IC1	SN76477	15	2
IC2	NE555	8	1

Figure 7: The Texas Instruments SN76477 is often used in a hardwired, dedicated device. One such use is simulated steam engine and whistle sound as shown here.

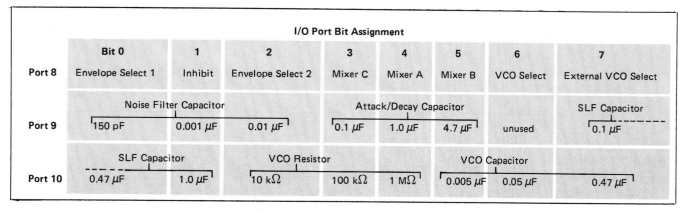

	Bit 0	1	2	3	4	5	6	7
I/O Port Bit Assignment								
Port 8	Envelope Select 1	Inhibit	Envelope Select 2	Mixer C	Mixer A	Mixer B	VCO Select	External VCO Select
Port 9	Noise Filter Capacitor 150 pF	0.001 µF	0.01 µF	Attack/Decay Capacitor 0.1 µF	1.0 µF	4.7 µF	unused	SLF Capacitor 0.1 µF
Port 10	SLF Capacitor 0.47 µF	1.0 µF	VCO Resistor 10 kΩ	100 kΩ	1 MΩ	VCO Capacitor 0.005 µF	0.05 µF	0.47 µF

Table 1: Designation of I/O (input/output) port assignments and associated component choices in the interface for the Texas Instruments SN76477 sound generator.

mately half the time. The voltage controlled oscillator is programmed to produce a whistle. Sufficient power to drive a speaker is facilitated by a two transistor complementary amplifier attached to pins 12 and 13.

Build a Computer Programmable Sound Generator Interface

While the SN76477 is not directly controllable by a computer as it exists, an interface between it and a computer can be designed which will give it some semblance of programmability. Figure 8 illustrates such an interface. Sound generation is programmed through three output ports, two of which control CMOS analog switches. These switches allow a variety of resistor and capacitor combinations to be selected. Total control requires three output commands from a BASIC or machine language program, and it is very easy to switch from a siren to a phaser gun sound when implemented as game sound effects. Photo 2 shows the video display of a typical spacewar game. Consider the sophistication that sound effects would add.

In the prototype, shown in photo 3, ports 8, 9, and 10 were chosen to drive the interface. Port 8 handles mixer and envelope selection; port 9 controls selection of components for the attack, decay and noise sections; and port 10 controls the SLF and VCO programming. The values chosen are nominal and will not allow unlimited sound synthesis. Potentiometers are added to facilitate fine tuning.

A More Sophisticated Programmable Sound Generator

The SN76477 is attached to a microcomputer largely through brute force. A far more sophisticated device has been

Sound Effect Desired	Hexadecimal Value Sent to Output Port		
	Port 8	Port 9	Port 10
Train	32	80	80
Phaser	B6	10	54
Siren	82	00	62

Table 2: Values which are sent to the output ports connected to the SN-76477 interface to produce the indicated sound effects.

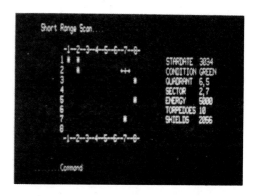

Photo 2: A typical video based space exploration game could be enhanced by sound effects.

Photo 3: A look at the prototype circuit of figure 7 attached to the back of an I/O (input/output) board.

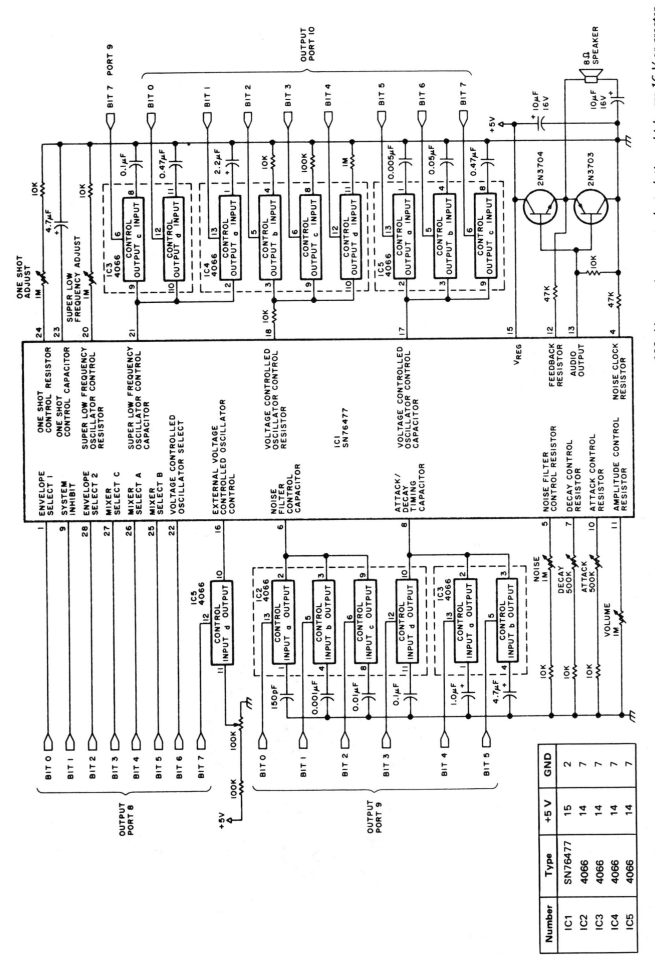

Figure 8: The SN76477 complete sound generator can be controlled by a computer. All capacitors are 100 V ceramic, except electrolytics which are 16 V or greater. All resistors are ¼W ±5%.

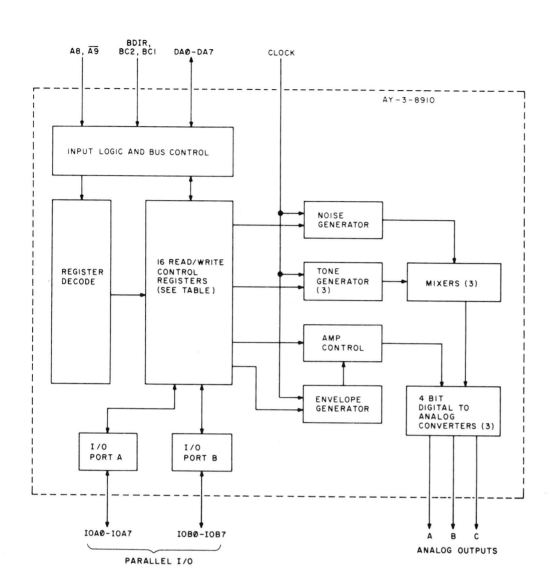

Figure 9a: Functional block diagram of the General Instrument AY-3-8910 programmable sound generator. The device is made by General Instrument Corp, Microelectronics Division, 600 W John St, Hicksville NY 11802.

REGISTER		B7	B6	B5	B4	B3	B2	BI	BO
RO	CHANNEL A TONE PERIOD	8 BIT FINE TUNE A							
RI		/////	/////	/////	/////	4 BIT COARSE TUNE A			
R2	CHANNEL B TONE PERIOD	8 BIT FINE TUNE B							
R3		/////	/////	/////	/////	4 BIT COARSE TUNE B			
R4	CHANNEL C TONE PERIOD	8 BIT FINE TUNE C							
R5		/////	/////	/////	/////	4 BIT COARSE TUNE C			
R6	NOISE PERIOD	/////	/////	/////	5 BIT PERIOD CONTROL				
R7	$\overline{ENABLE}$	$\overline{IN}$/OUT		$\overline{NOISE}$			$\overline{TONE}$		
		IOB	IOA	C	B	A	C	B	A
RIO	CHANNEL A AMPLITUDE	/////	/////	/////	M	L3	L2	LI	LO
RII	CHANNEL B AMPLITUDE	/////	/////	/////	M	L3	L2	LI	LO
RI2	CHANNEL C AMPLITUDE	/////	/////	/////	M	L3	L2	LI	LO
RI3	ENVELOPE PERIOD	8 BIT FINE TUNE E							
RI4		8 BIT COARSE TUNE E							
RI5	ENVELOPE SHAPE/CYCLE	/////	/////	/////	/////	CONT	ATT	ALT	HOLD
RI6	I/O PORT A DATA STORE	8 BIT PARALLEL I/O ON PORT A							
RI7	I/O PORT B DATA STORE	8 BIT PARALLEL I/O PORT B							

Figure 9b: Map of the control registers of the AY-3-8910.

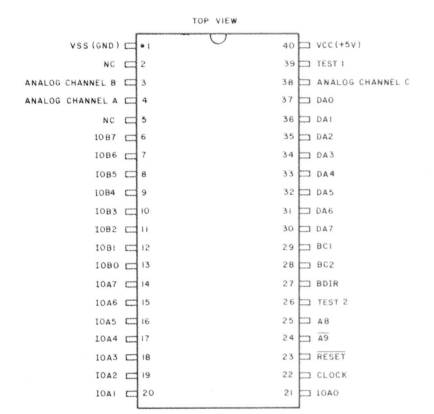

```
                          TOP VIEW

  VSS (GND) ▭ •1              40 ▭ VCC (+5V)
        NC ▭ 2               39 ▭ TEST 1
ANALOG CHANNEL B ▭ 3         38 ▭ ANALOG CHANNEL C
ANALOG CHANNEL A ▭ 4         37 ▭ DA0
        NC ▭ 5               36 ▭ DA1
      IOB7 ▭ 6               35 ▭ DA2
      IOB6 ▭ 7               34 ▭ DA3
      IOB5 ▭ 8               33 ▭ DA4
      IOB4 ▭ 9               32 ▭ DA5
      IOB3 ▭ 10              31 ▭ DA6
      IOB2 ▭ 11              30 ▭ DA7
      IOB1 ▭ 12              29 ▭ BC1
      IOB0 ▭ 13              28 ▭ BC2
      IOA7 ▭ 14              27 ▭ BDIR
      IOA6 ▭ 15              26 ▭ TEST 2
      IOA5 ▭ 16              25 ▭ A8
      IOA4 ▭ 17              24 ▭ A̅9̅
      IOA3 ▭ 18              23 ▭ RESET
      IOA2 ▭ 19              22 ▭ CLOCK
      IOA1 ▭ 20              21 ▭ IOA0
```

Figure 9c: Pin designations of the AY-3-8910 device.

recently introduced and it is designed specifically as a bus controlled device. This new device is the AY-3-8910 from General Instrument. It uses no external components and synthesizes sounds totally by digital means. A functional block diagram is shown in figure 9a.

You'll notice a similarity between this programmable sound generator and the Texas Instruments device in that they both contain the same elemental sound synthesis components such as noise and tone generators. The real difference is that the General Instrument programmable sound generator is programmed through 16 read/write control registers rather than resistors and capacitors. These registers appear as 16 sequential memory mapped I/O (input/output) locations to the controlling processor.

The AY-3-8910 incorporates a noise generator, three tone generators, three mixers, an envelope generator and three digital to analog converters for amplitude control. An added benefit is the inclusion of two decoded I/O ports which are available for other external applications. All subsystems are controlled through the control register array.

The device is specifically designed to interface with the General Instrument CP1600 series of microprocessors but it can be easily accommodated by others. Figure 10 illustrates this simple attachment. A bidirectional address/data bus, DA0 thru DA7, provides the necessary communication path. Since there are 16 registers, only four bits of address are actually used, and A8 and A̅9̅ serve more as device select lines by definition. BC1, BC2, and BDIR are the bus control lines and define bus direction, reading, and writing of register data. While an inexpensive circuit such as that shown in figure 11 can be used as the clock for both the processor and the programmable sound generator, they are basically independent and can be different rates. The programmable sound generator clock is primarily used for the sound synthesis. The reset line clears all registers.

For all practical purposes, signal line BC2 is unnecessary and can be tied to +5 V. The read/write control logic is shown in table 4.

The timing of BC1 and BDIR control lines are shown in figure 12. Data transfer is carried out by strobing these lines, while the address/data bus contains the pertinent contents. These pulses should be short and one processor clock cycle should suffice.

Whistling Bomb Sound Effect

Register Number	Hexadecimal Load Value	Explanation
Any not specified	00	
R7	3E	Enable tone only on channel A only.
R10	0F	Select maximum amplitude on channel A.
R0	30 (start)	Sweep effect for channel A tone period via a processor loop with approximately 25 ms wait time between each step from 30 to C0 (0.429 ms/2330 Hz to 1.72 ms/582 Hz).
R0	C0 (end)	
R6	0F	Set noise period to midvalue.
R7	07	Enable noise only on channels A,B,C.
R10	10	Select full amplitude range under direct control of envelope generator.
R11	10	
R12	10	
R14	10	Set envelope period to 0.586 seconds.
R15	00	Select envelope decay, one cycle only.

Phaser Sound Effect

Register Number	Hexadecimal Load Value	Explanation
Any not specified	00	
R7	3E	Enable tone only on channel A only.
R10	0F	Select maximum amplitude on channel A.
R0	30 (start)	Sweep effect for channel A tone period via a processor loop with approximately 3 ms wait time between each step from 30 to 70 (0.429 ms/2330 Hz to 1.0 ms/1000 Hz).
R0	70 (end)	
R10	00	Turn off channel A to end sound effect.

Table 3: Values which are loaded into the control registers of the General Instrument AY-3-8910 sound generator in order to produce the indicated sound effects.

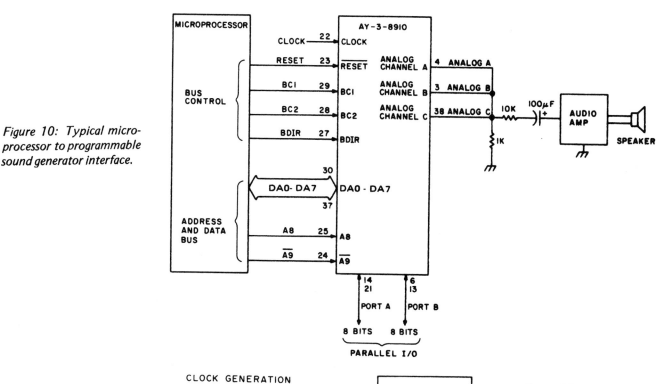

Figure 10: Typical micro-processor to programmable sound generator interface.

CLOCK GENERATION

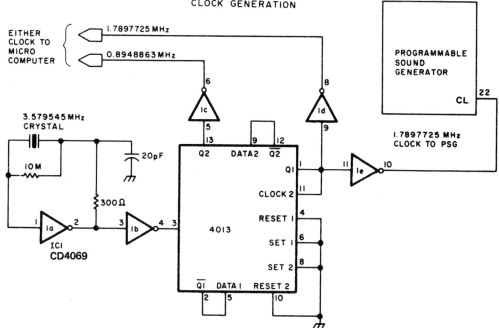

Figure 11a: A simple clock generator which can be used as a clock for the processor and the programmable sound generator.

AUDIO OUTPUT INTERFACE

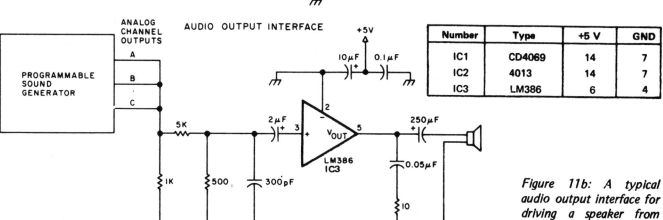

Number	Type	+5 V	GND
IC1	CD4069	14	7
IC2	4013	14	7
IC3	LM386	6	4

Figure 11b: A typical audio output interface for driving a speaker from the programmable sound generator.

BDIR	BC2	BC1	Function	CP1600 Function Abbreviation
0	1	0	Inactive	NACT
0	1	1	Read from PSG	DTB
1	1	0	Write to PSG	DWS
1	1	1	Latch Address	INTAK

Table 4: Summary of the read/write control logic needed to control the AY-3-8910 sound generator.

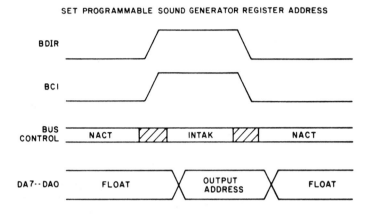

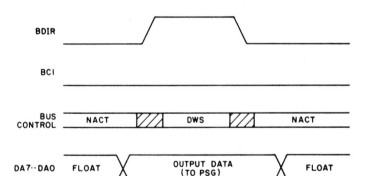

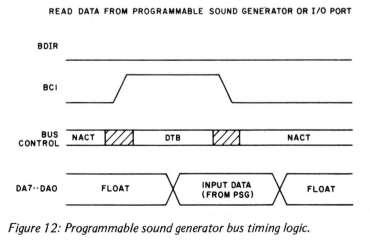

Figure 12: Programmable sound generator bus timing logic.

Tone Select

The registers are divided into six categories, and numbered in base eight:

Tone generators	R_0 thru R_5
Noise generator	R_6
Mixer control	R_7
Amplitude control	R_{10} thru R_{12}
Envelope control	R_{13} thru R_{15}
I/O ports	R_{16} and R_{17}.

Tones are square waves produced by dividing the input clock by 16, then counting that result down by a programmed 12 bit tone-period value. The 12 bit value, defined by the coarse and fine tune registers, is a combination of the two control registers. The 12 bits represent period and $T = 1/frequency$. The higher the register value, the lower the tone. Register contents range from 000000000001 (divide by 1) to 111111111111 (divide by 4095). With a 2 MHz clock the frequencies would be 125 kHz and 30.5 Hz respectively.

The other parameters, such as noise, mixers, amplitude, and envelope controls, are chosen in a similar manner. The actual programming technique is beyond the scope of this introduction to the AY-3-8910, and I suggest that interested readers send inquiries to General Instrument.

Connecting the AY-3-8910 to the S-100 Bus

Figure 13 shows how an AY-3-8910 programmable sound generator can be connected as an I/O device on the S-100 8080 compatible bus. Switches SW1 through SW6 define the starting I/O address of the 16 programmable sound generator registers.

Reading and writing from it is as illustrated.

Number	Type	+5 V	GND
IC1	7485	16	8
IC2	7485	16	8
IC3	7404	14	7
IC4	7402	14	7
IC5	7400	14	7
IC6	7400	14	7
IC7	74148	16	8
IC8	74LS367	16	8
IC9	74LS367	16	8
IC10	74LS367	16	8
IC11	74LS367	16	8
IC12	AY-3-8910	40	1

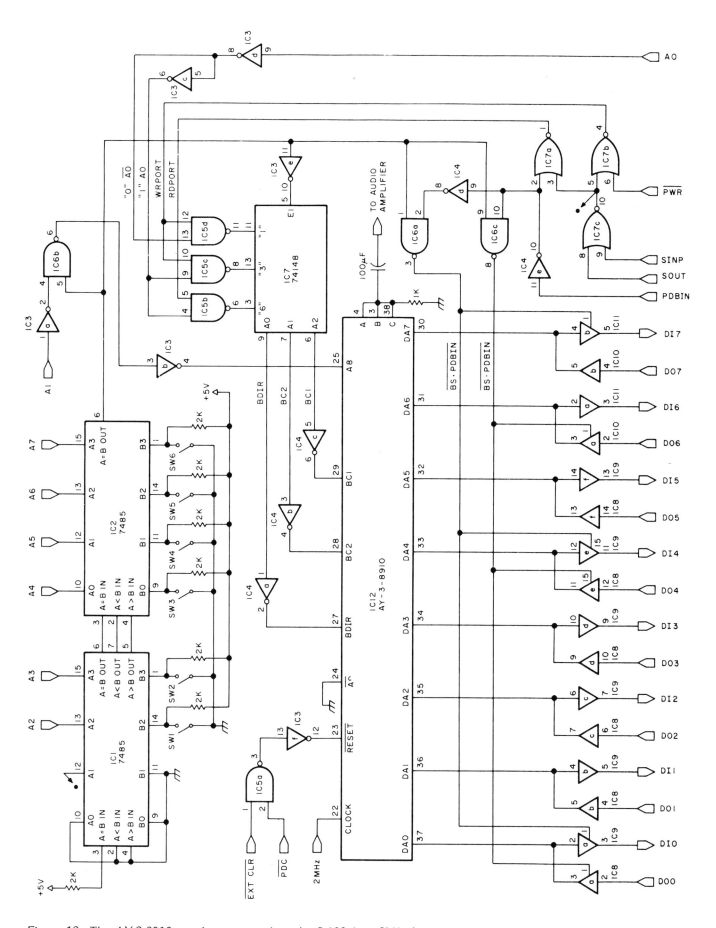

Figure 13: The AY-3-8910 can be connected to the S-100 bus. SW1 thru SW6 define the starting address of the 16 control registers. The power pin assignment is shown in the table at left.

LATCH ADDRESS ROUTINE

```
PORTADDR EQU 80H ; ADDRESS TRANSFER PORT ADDRESS
PORTDATA EQU 81H ; DATA TRANSFER PORT ADDRESS

; THIS ROUTINE WILL TRANSFER THE CONTENTS OF
; 8080 REGISTER C TO THE PSG ADDRESS REGISTER
PSGBAR    MOV  A,C ; GET C IN A FOR OUT
          OUT  PORTBAR ; SEND TO ADDRESS PORT
          RET
```

WRITE DATA ROUTINE

```
; ROUTINE TO WRITE THE CONTENTS OF 8080 REGISTER B
; TO THE PSG REGISTER SPECIFIED BY 8080 REGISTER C
;
PSGWRITE  CALL  PSGBAR ; GET ADDRESS LATCHED
          MOV   A,B ; GET VALUE IN A FOR TRANSFER
          OUT   PORTDATA ; PUT TO PSG REGISTER
          RET
```

READ DATA ROUTINE

```
; ROUTINE TO READ THE PSG REGISTER SPECIFIED
; BY THE 8080 REGISTER C AND RETURN THE DATA
; IN 8080 REGISTER B
;
PSGREAD   CALL  PSGBAR
          IN    PORTDATA ; GET REGISTER DATA
          MOV   B,A GET IN TRANSFER REGISTER
          RET
```

Listing 1: Routines written for the 8080 microprocessor to operate the AY-3-8910 programmable sound generator.

LATCH ADDRESS ROUTINE

```
; AT ENTRY, B HAS ADDRESS VALUE
;
LATCH CLRA
      STAA 8005 ; GET D DIR A
      LDAA #FF
      STAA 8004 ; OUTPUTS
      LDAA #4
      STAA 8005 ; GET PERIPHERAL A
      STAB 8004 ; FORM ADDR
      STAA 8006
      CLRA
      STAA 8006 ; LATCH ADDRESS
      RTS ; RETURN
```

WRITE DATA ROUTINE

```
; AT ENTRY, B HAD DATA VALUE
;
WRITE STAB 8004 ; FORM DATA
      LDAA #6 ; DWS
      STAA 8006
      CLRA
      STAA 8006 ; WRITE DATA
      RTS ; RETURN
```

READ DATA ROUTINE

```
; AFTER READ, B HAS READ DATA
;
READ STA A 8005 ; GET D DIR
     STA A 8004 ; INPUTS
     LDAA #4
     STA A 8005 ; GET PERIPHERAL
     DECA
     STA A 8006 ; READ MODE
     LDA B 8004 ; READ DATA
     CLRA
     STA A 8006 ; REMOVE READ MODE
     RTS        ; RETURN
```

Listing 2: Routines coded for the 6800 microprocessor to operate the AY-3-8910.

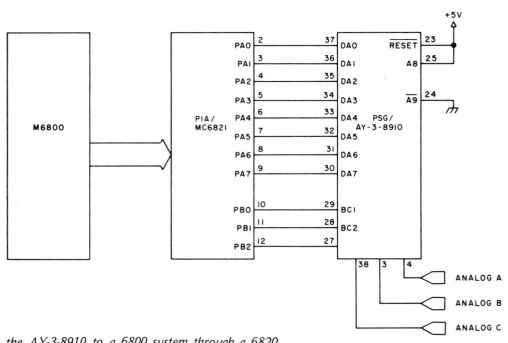

Figure 14: Connecting the AY-3-8910 to a 6800 system through a 6820 programmable interface adapter is easier than interfacing the S-100 bus.

A less complicated hardware interface is attained by using a peripheral interface adapter such as the 6820. Figure 14 demonstrates a technique which can be used for 6800 systems. The considerable difference in hardware complexity should in no way imply lack of ability using the 8080. If the S-100 bus is ignored and a 8255 programmable peripheral interface is used instead, it would result in a circuit similar to figure 14.

In Conclusion

I have briefly presented two methods of sound synthesis. While both are simple to implement, it is easy to recognize that the Texas Instruments part is more applicable in dedicated designs while the General Instrument device is for general synthesizer applications. It is not inconceivable that the AY-3-8910 could produce almost any sound, and it is a natural for use with a music interpreter running on a microcomputer. Perhaps the next famous composer will not direct a 150 piece orchestra but, rather, a trio of microcomputers controlling a bank of AY-3-8910s.∎

Circuit diagrams and drawings pertaining to the AY-3-8910 were provided courtesy of General Instrument Corp.

A BIT OF MUSIC

Dear Steve,

As a composer/performer, I found your article "Sound Off" quite intriguing. The potential of computer-controlled music generation inspired me to purchase a Commodore PET, but the tones generated by my rudimentary system are not exactly musical.

I would appreciate any improvements and suggestions you might have. The limiting factor in my case, and I am sure this is true for others, is lack of proficiency with the instrument.

Jack Hobson

My talents are geared more toward building the instrument than making music on it. If you are reasonably adept at building circuitry, there is a way to run the General Instrument AY-3-8910 Programmable Complex Sound Generator from the parallel user port (J2) of the PET. The clocking of the integrated circuit is not critical, only the sequence of events, but the circuit does require 11 bits of information.

The circuit of figure 1 here uses only 74LS95 4-bit parallel-access shift registers. IC1, IC2, and IC3 are paralleled to form a 12-bit shift register, and IC4, IC5, and IC6 make up a 12-bit latch. By setting the appropriate logic level on bit 7 of the user port, the information will be loaded into the 12-bit register when a high/low/high transition occurs on bit 1. When 12 bits have been loaded, a low/high/low transition on bit 0 can be used to latch the binary value and stabilize the information while more is loaded.

*The fast action of the shift and store operations should be fairly transparent to the AY-3-8910, which should operate as described in the article....***Steve**

A BIT MORE MUSIC

Dear Steve,

I read with interest your article on the AY-3-9810, and I am presently building the interface for my Southwest Technical Products 6800 system. Do you intend to publish any software for the 6800-based processors that will drive the circuit?

Arnold Pung

The AY-3-9810 is made by General Instrument on Long Island (600 W John St, Hicksville NY 11802). They are the people to contact about particular applications of that device. If you write or call (516) 733-3107, direct your inquiry to the product manager associated with the circuit.

*It is important to talk to the right person since one manager may not be familiar with another's product line. As most companies do, General Instrument Corporation puts out a large amount of application literature....***Steve**

Anyone Know the Real Time?

I'm sure you've all heard the term *real-time*, such as a real-time operating system. But, how many really understand its meaning? A simple definition of a real-time system is: a system that operates in real time, that is, it responds to the need for action in a period of time proportional to the urgency of the need; first things are done first. In control applications the system can be depended on to provide the information necessary to base time-dependent decisions on information that is up to date as of the minute or the hour. Real time describes the processing of information in a sufficiently rapid manner that the results of the processing are immediately available to in-fluence control of the process being monitored.

While there are particular architectural enhancements in high-speed process monitoring and control systems, basically any computer can be configured to perform some semblance of real-time operations. The essential criterion is that the computer be capable of performing a specific action at a particular time. The extent of real-time operation then becomes dependent upon execution speed. If a program that takes 1 second to analyze a data input and display it on the video display is to run in real time, it can only be called once per second. For continuous sampling this also means that the

computer cannot be tied up doing any other task without provision being made for that program to be interrupted so that the analysis program can run. Most often, computers utilize hardware priority interrupts to provide this capability. A direct benefit of this approach is that all programs can execute asynchronously, since interrupt logic synchronizes the computer's action upon the occurrence of a real-time event. Further discussion of interrupts will continue later in this article.

A second, slightly less complex method of synchronizing computers to real-time events is through a technique of *status scanning* (or device polling). This software-intensive situation requires that all devices demanding real-time interaction set status flags to indicate ready conditions. The computer scans these flags periodically and performs the appropriate action. The flags are reset when the devices have been serviced. It is important to keep in mind that all the programs that the computer normally executes must be short enough to allow the computer to service every device. Also, care must be taken to design the system so that a second event cannot occur on an individual device before the computer has acknowledged the first event.

Most sophisticated real-time systems use a combination of these two methods. A clock circuit, such as that in figure 1, provides a time "tick" to the processor's nonmaskable interrupt line. This can be every 60th, 10th, or 1 second, as suggested in this schematic. When the computer acknowledges the interrupt, it first saves all registers from the program it

Photo 1: A prototype board of real-time clock mounted on the back of an existing parallel I/O (input/output) board. Two reed switches on the left side of the board are for manual setting of the clock. The empty sockets are used for the particular application for which this board was designed, a home security system.

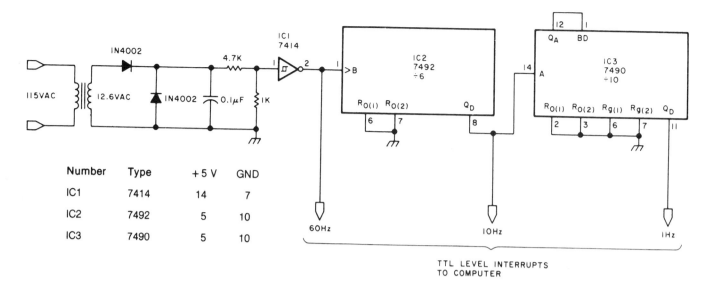

Number	Type	+5 V	GND
IC1	7414	14	7
IC2	7492	5	10
IC3	7490	5	10

Figure 1: A simple time-base generator for an interrupt-driven real-time clock.

was executing, and then services the real-time interrupt. Frequently the first action is to increment an internal counter which keeps track of elapsed time. Usually it will be a value equivalent to the total number of clock ticks, whether in seconds or milliseconds. Once this regular interval has been established, it is easy for the computer to scan all status flags from real-time devices. The addition of more real-time activities for the processor does not entail multiplying the number of interrupt lines, but rather it simply entails placing another status flag on the list of those to be checked on each clock tick.

The choice of a totally interrupt-driven real-time system, a combination scan and interrupt type, or a total scanning system is dependent upon the quantity of real-time operations and their frequency. An interrupt-driven system can process information faster than the same system configured for real-time scanning.

Real-Time Applications for Personal Computers

So far I have emphasized the system attributes, but nowhere have I discussed applications, particularly personal computing applications. Clock divisions down to milliseconds sound great and make interval timing extremely accurate, but I doubt that the majority of home computerists would want something that complex to integrate into their system. If my mail is any indication of this, they would prefer the design of a real-time

clock which can be directly applied in home control applications. Automatically turning on the percolator at 6:45 AM would be far more stimulating than a high-speed data acquisition system which few would need.

Build a Real-Time Clock

Essentially, the kind of real-time system which might appeal to personal computer users is one with a resolution of perhaps 1 minute rather than 1 ms. It should be read directly in hours and minutes like a 4- or 6-digit clock and not just total counts.

A direct benefit of low resolution is reduced overhead. The computer does not have to acknowledge the clock update or scan status flags as often. It may not seem like much of a time savings, considering instruction execution speeds of 1 μs. However, the interrupt routine could be 30 bytes and 100 μs long. If called every millisecond it would eat up 10% of the total cycle time—just to increment a counter! When it comes to real time, be careful not to byte (sic) off more than you can process.

The easiest way to provide an

DEVICE	ACTIVATION	DEACTIVATION	PRESENT STATE
1. Night Light	2000	0130	ON
2. Driveway flood	1930	2230	OFF
3. Coffee Perk	0645	0730	OFF
4. Water Softener	0230	0430	OFF
5. Outside Lights	2000	2330	OFF
6. Thermostat Dn	2300	0530	ON
7. Bedroom TV	0700	0900	OFF
8. Dehumidifier	0300	1800	OFF

PRESENT TIME ---23 Hours 47 Minutes---

SYSTEM STATUS ****** GREEN ******

Photo 2: A typical application of a real-time clock. This display is from my computer-controlled security system.

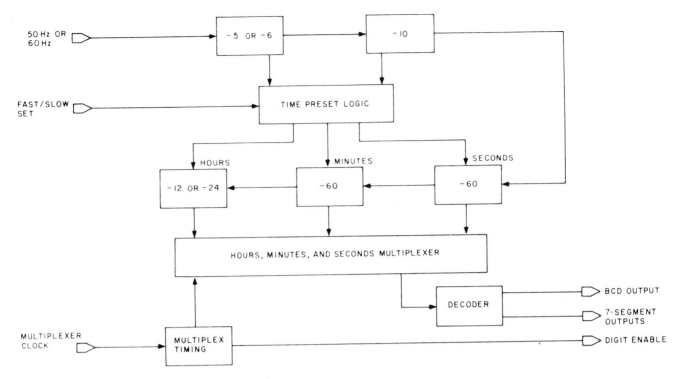

Figure 2: The block diagram for a typical clock chip.

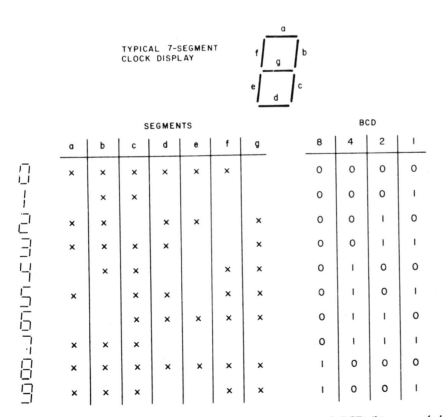

| | SEGMENTS | | | | | | | BCD | | | |
	a	b	c	d	e	f	g	8	4	2	1
	✗	✗	✗	✗	✗	✗		O	O	O	O
		✗	✗					O	O	O	I
	✗	✗		✗	✗		✗	O	O	I	O
	✗	✗	✗	✗			✗	O	O	I	I
		✗	✗			✗	✗	O	I	O	O
	✗		✗	✗		✗	✗	O	I	O	I
			✗	✗	✗	✗	✗	O	I	I	O
	✗	✗	✗					O	I	I	I
	✗	✗	✗	✗	✗	✗	✗	I	O	O	O
	✗	✗	✗			✗	✗	I	O	O	I

Figure 3: A comparison of output codes from 7-segment and BCD (binary coded decimal) clock chips.

hourly and minute by minute input is to interface the computer to an MOS/LSI (metal oxide semiconductor/large scale integrated) clock device such as that found in most digital clocks or watches. The block diagram of a typical clock chip is shown in figure 2. This LSI device replaces about 22 TTL (transistor-transistor logic) chips once necessary to perform the same function, and consumes very little power, allowing battery standby operation. The circuit of figure 1 uses inexpensive TTL rather than CMOS (complementary metal oxide semiconductor) because battery backup is irrelevant if the computer cannot acknowledge interrupts in a powered down state. Figure 3 illustrates the logic of the BCD (binary coded decimal) and 7-segment output lines.

There are two approaches to the design of a clock interface. One approach is to let the clock circuit operate independently from the computer, attached in such a way that the computer is able to monitor this activity and extract a time value. The second approach, which I prefer, is to give the computer complete control

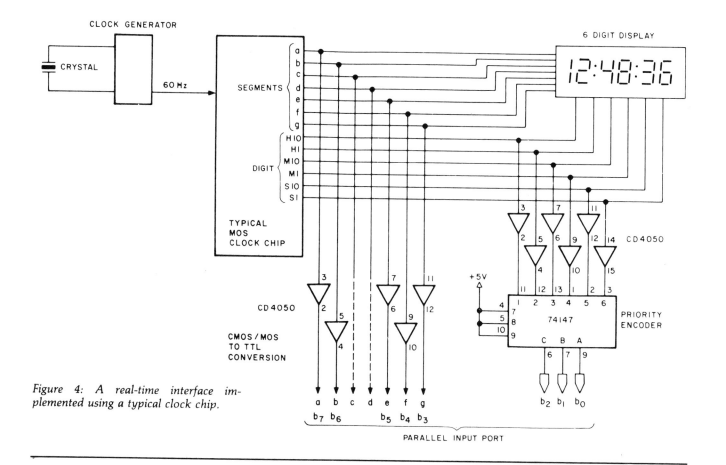

Figure 4: A real-time interface implemented using a typical clock chip.

over the information flow of the clock in a synchronous manner. This design makes the interface speed independent and allows it to be used directly with high-level languages.

Figure 4 shows the typical real-time clock interface. In this design the clock is configured in the usual manner to drive a 6-digit light emitting diode display. The clock runs independently with the display multiplexing rate (about 1 kHz) set by a resistor/capacitor combination attached to the chip. Five of the 7-segment drive lines are level shifted and buffered for TTL through a CD 4050, and the 6 digit lines are priority encoded to produce a 3-bit binary value for transmission to the computer of the energized digit-enable line. The 3-bit digit and 5-bit segment codes are combined to produce a single 8-bit byte interfaced to a parallel input port.

In operation, the computer program first looks at bits b_0 thru b_2 to determine which digit of the display to activate. Then it reads bits b_3 thru b_7 and compares them to a table to determine which character is being displayed. (Only 5 of the 7 segments

are necessary to perform this comparison.) This process is repeated 5 more times as the chip sequences through the other digits. The final result is formatted into hours, minutes, and seconds. The entire operation takes about 10 ms and requires that the program be written in machine language.

If you can believe the claims of the manufacturers, there are now more computers in use that run BASIC rather than machine language as their primary mode of interactive communication. While it is still possible to manipulate individual bits and write machine language device control subroutines for these computers, their owners are obviously more familiar with high-level languages and would necessarily feel more comfortable with a clock design which could be controlled in BASIC as well as machine code. Figure 5 demonstrates such a design.

This circuit, which can be manually or automatically preset, is fully static and allows the display output lines to be completely under program control. The basic 5-chip interface consists of a 4-digit BCD/7-segment

output clock type MM5312, an MM5369 time-base generator, 2 MOS to TTL buffers to send data to the microprocessor, and 1 TTL-to-CMOS converter for processor control over the clock chip. Time is read by the computer as 4 binary coded decimal numbers. In a 4-digit clock like the one in figure 5, the data appears as a digit-enable output and an associated BCD value. The tens of minutes data is available when bit b_5 is high (bits b_4, b_6, and b_7 are low). It will appear as a BCD quantity in bits b_0 thru b_3. Unlike the circuit of figure 4, this unit is static and has no display to drive. It will stay on a particular digit until it is instructed to sequence to the next digit. This is accomplished by controlling the display-multiplexer input line of the clock.

Figure 6 shows how the multiplexer line is controlled in this application. Bit 0 of an output port (port 8 in my example) is used to pulse multiplexer input pin 22. At any time, 1 of the 4 digit-enable output lines will be low (at the chip), indicating that the multiplexer is set on that digit. The data on the BCD lines is for that digit. Reading the next digit is simply a case

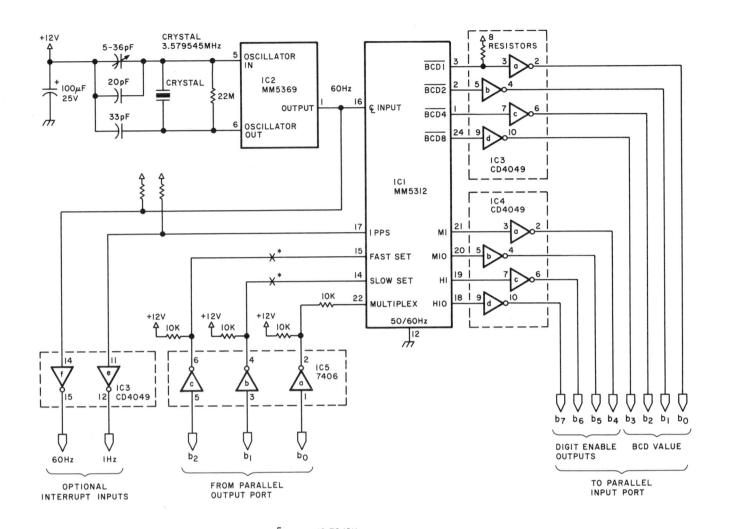

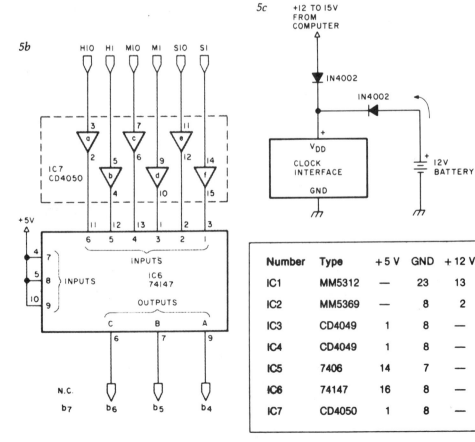

Figure 5: Design for a real-time clock which can be synchronously controlled by a BASIC or machine language routine. 5a shows the schematic diagram; asterisks indicate lines which should be opened to prevent loss of time data when the computer is powered down and the interface is used with battery backup. 5b shows an alternate configuration for a 6-digit clock when using an MM5311 integrated circuit. 5c shows the circuit for battery backup operation. The clock interface requires 12 mA from the battery during standby (indicated by the arrow).

Number	Type	+5 V	GND	+12 V
IC1	MM5312	—	23	13
IC2	MM5369	—	8	2
IC3	CD4049	1	8	—
IC4	CD4049	1	8	—
IC5	7406	14	7	—
IC6	74147	16	8	—
IC7	CD4050	1	8	—

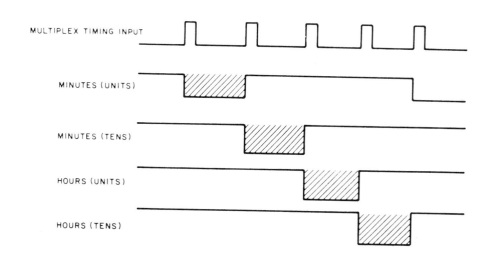

Figure 6: Display multiplex timing sequence for the circuit in figure 5.

MULTIPLEX TIMING INPUT

MINUTES (UNITS)

MINUTES (TENS)

HOURS (UNITS)

HOURS (TENS)

INDICATES THAT BCD LINES CONTAIN VALID DATA FOR RESPECTIVE DIGIT DURING THIS PERIOD (AFTER APPROXIMATELY 200 μSEC SETTLING TIME)

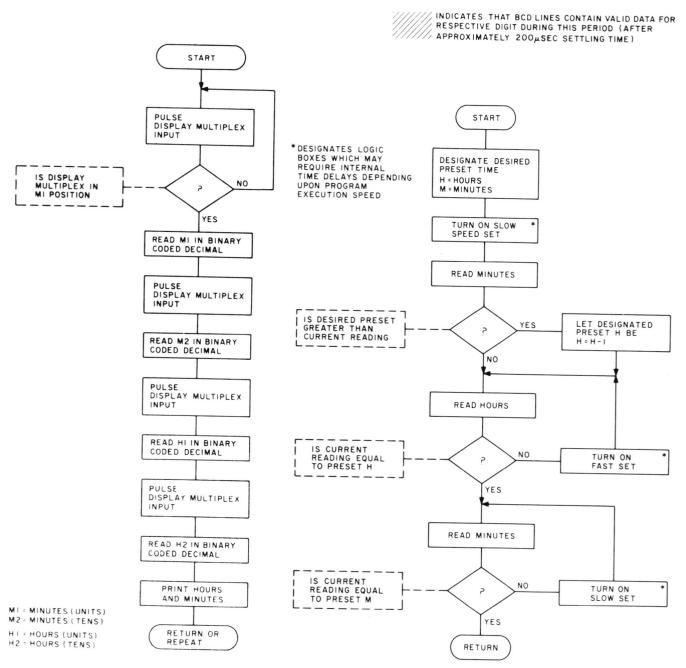

M1 = MINUTES (UNITS)
M2 = MINUTES (TENS)

H1 = HOURS (UNITS)
H2 = HOURS (TENS)

Figure 7: Flowchart of the program given in listing 1.

Figure 8: Flowchart for the automatic reset routine.

Listing 1: Program for the real-time clock.

```
LIST

100 REM REAL TIME CLOCK
110 REM COPYRIGHT 1979 STEVEN CIARCIA
120 REM THIS SIMPLE PROGRAM ALLOWS A COMPUTER TO TELL TIME BY
130 REM INTERFACING A DIGITAL CLOCK CHIP TO AN I/O PORT. (PORT 8 IN THIS EXAMPLE)
140 REM THE DISPLAY MUX LINE IS CONTROLLED BY THE COMPUTER. FIRST IT IS PULSED UNTIL
150 REM IT IS SET ON THE LEAST SIGNIFICANT DIGIT
160 OUT 8,1 :OUT 8,0 : T=INP(8) :D=T AND 16
170 IF D=16 THEN 200 ELSE 160
180 REM ONCE THE LSD POSITION IS SET THE 4 SUCESSIVE READINGS ARE TAKEN
190 REM THE INPUT PORT DATA IS ANDED WITH 15 TO OBTAIN THE BCD VALUE (REMEMBER,BASIC USES
    DECIMAL)
200 M1=T AND 15 :GOSUB 250 :REM    MINUTES (UNITS)
210 M2=T AND 15 :GOSUB 250 :REM    MINUTES (TENS)
220 H1=T AND 15 :GOSUB 250 :REM    HOURS (UNITS)
230 H2=T AND 15 :GOSUB 250 :REM    HOURS (TENS)
240 PRINT H2;H1;":";M2;M1 :GOTO 160
250 OUT 8,1 :OUT 8,0 :T=INP(8):RETURN :REM ADVANCE DISPLAY MUX

READY
```

of pulsing bit b_0 again. There is no time constraint either. You can wait 10 minutes between digits if you wish (but the data won't mean much). It is best to read the 4 digits sequentially. The circuit is easily interfaced and exercised in BASIC as demonstrated in listing 1. The flow diagram of this program is shown in figure 7.

The addition of 2 more gates connected to output bits b_1 and b_2 facilitate automatic time preset. Figure 8 follows the logic of how such a program could be written. Two magnetic reed switches shown in photo 1 can be attached between pins 14 and 15, respectively, and ground to allow manual preset as well. I find that it is easier to just turn on the clock program in continuous display mode and adjust the clock as I read it. If a battery back-up capability is added, the 2 TTL automatic set gates should be disconnected. When the computer is powered up, random data can appear on bits b_1 and b_2, accidently causing it to enter the set mode. This is not a problem on the input. While a 4-digit, 24-hour clock is quite enough in my application (an example is shown in photo 2), there are those who need a second designation. Substituting an MM5311, the s_1 and s_{10} digit-enable line can be added as 2 more parallel input bits and treated exactly as the present circuit, or binary encoded to reduce input bits, as shown in figure 5b. This method will require a slight software change but should be an equally viable approach. The present program in listing 1 executes in approximately 50 ms when used with Micro Com 8 K Zapple BASIC, but it works equally well with a machine language routine.

Whatever your final configuration, I am sure you will find that accurately timed control outputs are a definite advantage on any system. And there is no reason for the hardware of any interface to constrain the operator's choice of software interaction if it is not dictated by the frequency of events themselves. ∎

Joystick Interfaces

Photo 1: A typical joystick with 4 potentiometers.

Photo 2: Note how moving the stick moves the gimbal arrangement, which in turn changes the settings of the potentiometers.

The thought that often comes to mind when the word joystick is mentioned to a computer enthusiast is of a spacewar-type game. A photon torpedo is fired from an opponent's starship, and the thruster joystick is deftly moved to reposition the craft out of its path. All of this occurs without having to take your eyes off the screen. Eye/hand coordination is almost "instinctive." With a glance to the upper right of the video screen, the joystick is tilted to the upper-right corner of its 360° range. This moves the spacecraft toward that coordinate. Reverse thrust is accomplished by moving the joystick in the opposite direction, as though you are pulling back on the throttle of a real

craft. Such is the general experience with joysticks. However, the potential use of these devices greatly exceeds that of game playing.

A joystick, for those people who are unfamiliar with one, is shown in photo 1. It is an electromechanical device with resistance outputs proportional to the X,Y displacement of a central ball and lever. Photo 2 illustrates the mechanical connections to the potentiometers.

When the stick is positioned in the center of its axes, the X and Y potentiometers show resistances in the center of their ranges. When the stick is tilted to the upper right, both potentiometers are at their full-resistance limit, while the opposite

(lowest resistance) is true when in the lower-left position. The outputs of the 2 potentiometers accurately track, as if on an X,Y coordinate axis, the position of the joystick. It should be noted that while it takes only 2 potentiometers to define 2-dimensional travel, most joysticks are manufactured with 4 potentiometers. This is a remnant of the days when joysticks were connected directly to the 4 deflection-plates of a cathode ray tube (video screen).

It is one thing to *consider* interfacing a joystick to a computer, and quite another to *do* it. A joystick is a mechanical X,Y positioning device. Even with proportional output resistances, an input interface must be designed to convert position from an analog to a digital representation which can be used by the computer. A further consideration is the resolution, or percent, of full-scale travel per bit sensitivity. Is the application so gross that center and full-scale are the only points of interest, as in a

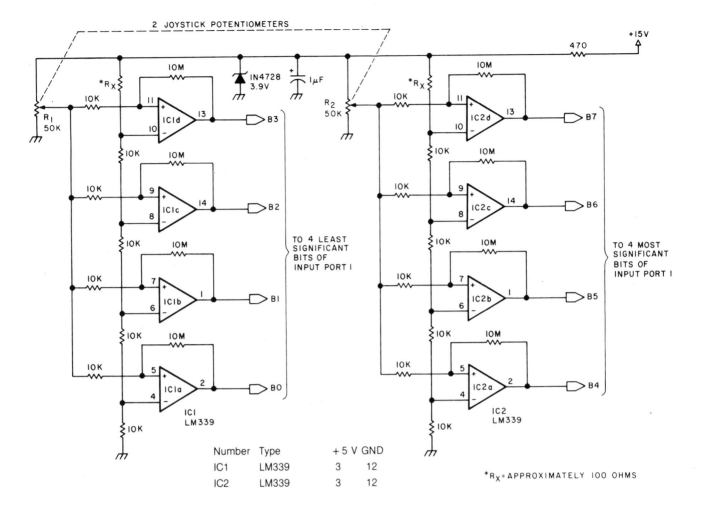

Figure 1: *Low-resolution static interface. This interface is for a 2-potentiometer joystick. For 4-potentiometer joysticks, build a second circuit like this one, and interface it to another input port. Note that if the comparator does not trigger at full-scale setting, a small resistor may have to be added at Rx (marked with asterisk).*

game control, or is the application one which requires fine control, such as a cursor-positioning device in a high-resolution graphics system?

All joystick interfaces are not created equal. There is a trade-off between hardware and software. The lower the resolution, the fewer the parts. The higher the resolution, the greater the electrical complexity or the software interaction with the interface. It is also important to recognize that computer systems which operate *only* in a high-level language like BASIC cannot use an interface design that requires an assembly-language subroutine as an integral component. In such instances only a static interface can be used.

Included in this presentation are 4 interface designs which should cover most requirements, as well as demonstrate the considerable differences between them. The 4 types are:

- low-resolution static
- high-resolution fully static hardware
- software-driven pulse-width modulated
- high-resolution analog-to-digital

Low-Resolution Static Interface

First of all, *static* simply means that the interface hardware determines the potentiometer position value and presents it in constant, parallel digital form to the computer. When the interface is attached to any parallel input port, this joystick value can be read with a single INPUT command in BASIC. As far as the computer is concerned, the value is fully static, and the computer reads whatever data is there when the INPUT is executed. The interface hardware has the responsibility of asynchronously updating the digital value as the stick is moved.

Often the joystick is simply used to indicate relative direction and magnitude. In a wheelchair, for instance, full linear control of speed and direction would require rather expensive drive electronics. Most chairs use simple relay contacts and provide 2 or 3 selectable speeds. A joystick control built for this application would not have to have a resolution of 8 bits, but could, in fact, suffice with 2. Figure 1 shows a low-resolution static output joystick interface suitable for use in this application.

Each potentiometer is connected as a voltage divider between a reference voltage source of 3.9 V and ground. The voltage output of each potentiometer is, in turn, fed to a 2-bit, parallel analog-to-digital converter. This type of converter uses 4 comparators set for 25%, 50%, 75%, and 100% of full scale. If a voltage, when applied, is less than 0.975 V, all comparator outputs will be at 0 V. At 1.0

```
              MVI    B         clear B
              OUT    FF,0      trigger one-shots
    AGAIN     INR    B         increment B register
              IN     FF        read potentiometers
              ANA    01        isolate bit 0
              JNZ    AGAIN     continue as long as one-shot is high
              HLT              value is in B register
```

Listing 1: A typical assembly-language program for using the joystick interface of figure 4. After the one-shots are triggered, the program loops and checks the status of bit 0. When this bit is set, the conversion value is in register B. This program assumes that there is only 1 value being checked, and it is being input through bit 0.

V, corresponding to the joystick being moved 25% of full scale, the least-significant bit (LSB) of the converter will be a logic 1, while the other bits are low. Similarly, at full input all comparators will be triggered, and bits 0 thru 3 will be logical 1s.

Additional encoding logic can be added to produce a true 2-bit representation from the 4 comparators, but it is just as easy for a computer to interpret it directly. With a 4-bit connection as shown, used in a BASIC program, 25% of full scale would be 1 decimal, 50% of full scale would be 3 decimal, 75% of full scale would be 7 decimal, and full scale would be 15 decimal. It should be easy to trigger any action by a coincidence with these values. The real significance of this method is that the potentiometer position is presented statically to the computer and requires no other interaction. This makes it ideal for direct use with BASIC.

High-Resolution Static Interface

It is quite possible that 2 bits of resolution is not enough for your application, but direct compatibility with a slow, high-level language is still a requirement. Expanding the parallel comparator method will work in theory, but you must realize that a 4-bit analog-to-digital converter uses 15 comparators, and an 8-bit, parallel analog-to-digital converter needs 255 comparators! So much for that method.

Realizing that the output of the joystick is a variable resistance, we can use this to advantage. This resistance can set the *time constant* of a function which has a pulse width proportional to joystick position. Figure 2 illustrates an interface design which uses this technique.

The 2 joystick potentiometers R1 and R2 control the pulse width of a one-shot (monostable multivibrator). The one-shot has a pulse width of 35 ms when the potentiometer is at 50 k ohm full scale and something less than 100 μs at 0% of full scale. A 7.5 kHz clock signal asynchronously triggers the one-shots. When the one-shot fires, its duration is proportional to the joystick position and will vary from approximately 0 to 35 ms. Using midscale pulse width of 17 ms as an example, the circuit timing is as in figure 3.

On the leading edge of the one-shot signal, a *clear* pulse is generated through an edge detector configured 7486 device. The clear pulse resets the **two 7493s which form an 8-bit counter. Once cleared, the counters start counting clock pulses for the duration of the one-shot's period. On its trailing edge, a *load* pulse is generated which loads this 8-bit counter into an 8-bit storage register. The computer is connected to read this 8-bit value through a parallel input port. Successive clearing and counting operations update the register every 35 ms or so (worst case). The clock rate is 7.5 kHz which has a period of 133 μs. If the one-shot has a pulse width of 17 ms, the 127 clock pulses would be gated to the counter. Of a total possible 255 counts, 127 would represent 50% of full scale.**

Software-Driven Interfaces

So far I have discussed only static interfaces. If the computer used with the joystick has sufficient speed and excess computing time available, then it is reasonable to use the computer to directly determine the one-shot period.

Figure 4 shows a circuit which

directly connects to the computer bus and demonstrates this technique. The circuit as shown is wired for I/O (input/output) port decimal 255 or hexadecimal FF. The 4 joystick potentiometers are used as the timing resistors on 4 NE555-type one-shots. When an OUT 0, FF is executed in assembly language, it triggers all 4 one-shots. To keep track of the pulse widths, a 74125 3-state driver gates the one-shot outputs onto the data bus during an IN FF instruction. By looping through this program a number of times and keeping track of the logic levels of the 4 one-shots, the computer can accurately determine joystick position in terms of loop counts of instruction times. Listing 1 is a program which does this for 1 potentiometer.

High-Resolution Analog-to-Digital

While all methods are in *some* way analog-to-digital converters, the last method is in fact an 8-bit absolute-analog-to-digital converter, typical of the type used in computerized measurement applications. IC1 is an 8-bit digital-to-analog converter that produces an output voltage proportional to a digital input applied to pins 5 thru 12. For a complete explanation of this device, I refer you to a previous "Ciarcia's Circuit Cellar" article, "Control the World" (September 1977 BYTE, page 30). This article also outlines calibration and test procedures.

The 3 basic sections are a computer-controlled voltage source (ICs 1 and 2), an analog-input multiplexer (IC3) which selects an individual joystick potentiometer by a 2-bit address code, and a comparator (IC4) which compares these voltages. In operation, the digital-to-analog converter is first set to 0 V out (hexadecimal 00 digital input to it) and 1 potentiometer is selected through the multiplexer. If V0 from the digital-to-analog converter is less than V_{in} from the potentiometer, the output will be logic 0. Next, the digital-to-analog converter input setting is incremented, and the comparator output is checked again.

Eventually an input count will be reached which will exceed V_{in}. The comparator output will then be a logic 1. The digital-to-analog converter input count is now the value of the voltage V_{in}. The worst case requires 256 iterations using this

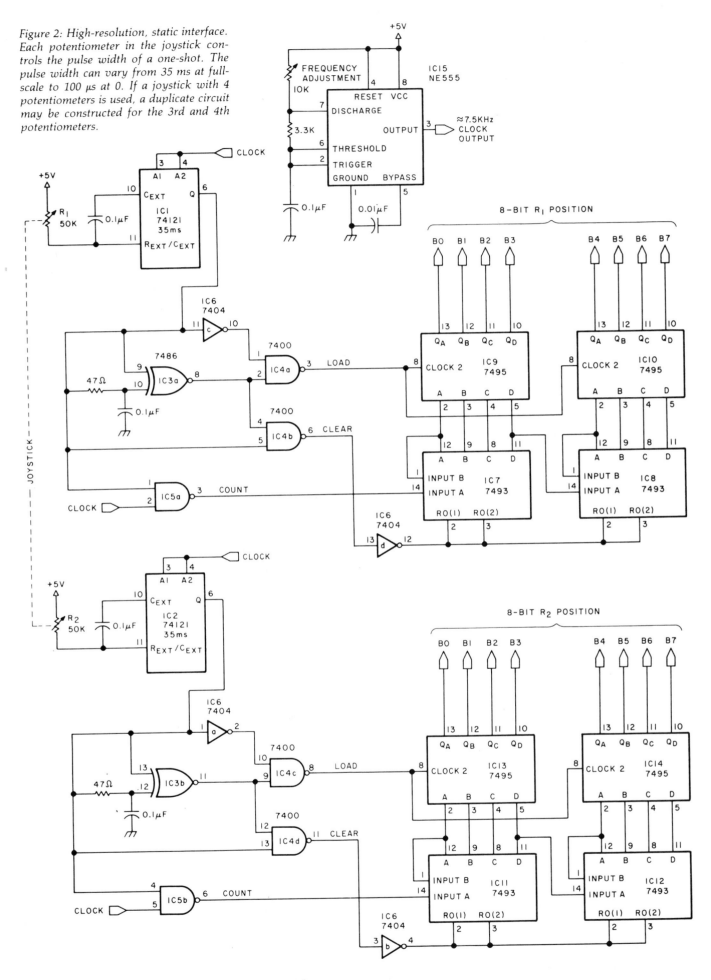

Figure 2: High-resolution, static interface. Each potentiometer in the joystick controls the pulse width of a one-shot. The pulse width can vary from 35 ms at full-scale to 100 μs at 0. If a joystick with 4 potentiometers is used, a duplicate circuit may be constructed for the 3rd and 4th potentiometers.

Number	Type	+5 V	GND
IC1	74121	14	7
IC2	74121	14	7
IC3	7486	14	7
IC4	7400	14	7
IC5	7400	14	7
IC6	7404	14	7
IC7	7493	5	10
IC8	7493	5	10
IC9	7495	14	7
IC10	7495	14	7
IC11	7493	5	10
IC12	7493	5	10
IC13	7495	14	7
IC14	7495	14	7
IC15	NE555	8	1

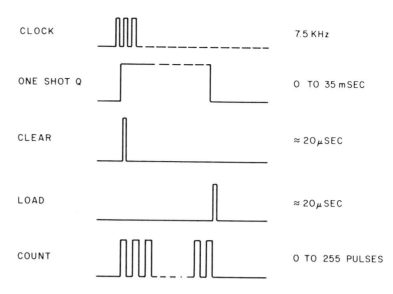

Figure 3: Timing diagram for interface of figure 2. The driving clock signal is 7.5 kHz. The one-shot can be triggered for periods of 0 to 35 ms, depending upon the position of the joystick. When a reading is to be taken, the counters are cleared. Counts are made until the one-shot signal drops, and then a load signal is sent to the interface. At this point the counter is read to determine the position of the joystick.

Number	Type	+5 V	GND
IC1	NE556	14	7
IC2	NE556	14	7
IC3	7430	14	7
IC4	7400	14	7
IC5	74125	14	7

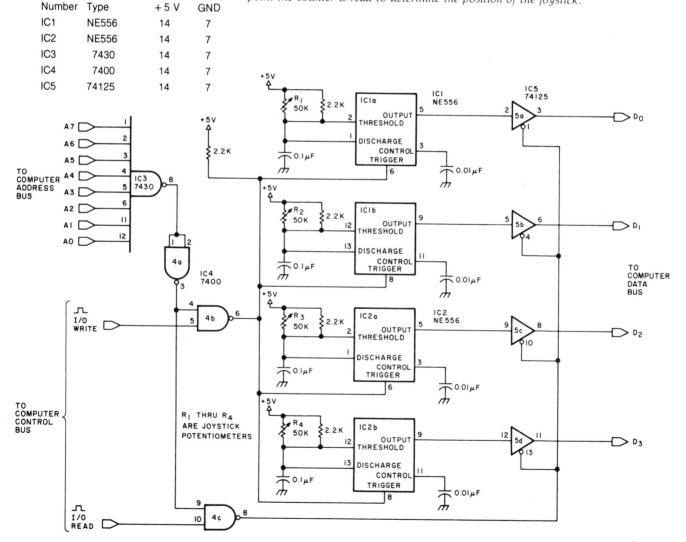

Figure 4: Software-driven interface. If the computer can directly read the input from the joystick interface, the hardware required can be greatly simplified. When hexadecimal FF is output to port 0, all 4 one-shots are triggered. The pulse width is then determined by a program running through a short loop looking at the logic levels of the 4 one-shots. Listing 1 shows a typical program for this application.

Number	Type	+5V	GND	+15V	+15V
IC1	MC1408-L8	13	2	3	
IC2	LM301A			4	7
IC3	CD4051	4			8
IC4	LM301A			4	7

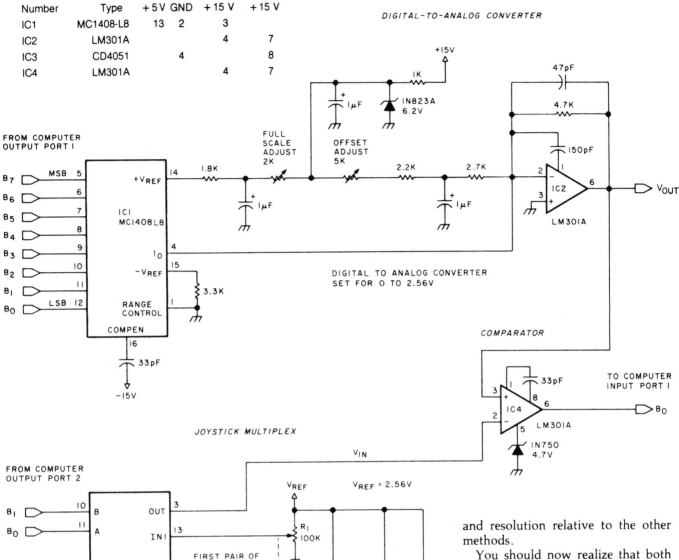

Figure 5: High-resolution analog-to-digital conversion. This hardware-oriented device multiplexes 4 voltage inputs (from the joystick potentiometers) and has the capability of handling 4 more voltages.

method. A better technique is successive approximation where the computer progresses through a binary search to "zero in" on the final value. A full explanation of successive approximation is delineated in my article entitled "Talk to Me: Add a Voice to Your Computer for $35" (June 1978 BYTE, page 142).

With the digital-to-analog converter set for a full-scale value of 2.56 V, each count is equivalent to 10 mV. Only 4 channels of the CD4051 are used for the joysticks, leaving another 4 channels as auxiliary inputs from external sources. Thus it is possible for this interface to serve a dual role because of its high accuracy

and resolution relative to the other methods.

You should now realize that both the design and construction of a joystick interface are influenced by many factors. It is not unusual to find one manufacturer charging $50 for a joystick, while another charges $200. Resolution, accuracy, and software interaction are the prime considerations. Where static inputs are required, the hardware will necessarily be more complicated. Resolution and accuracy ultimately determine the complexity of the interface.

For simple spacewar-type games, the circuit of figure 1 should suffice. For more demanding applications such as cursor control in a high-resolution graphics system, figure 5 may be the optimum choice. Be careful when buying joystick interfaces. Make sure that they mate with your program requirements and your system's abilities. ∎

Self-Refreshing LED Graphics Display

Light emitting diodes (LEDs) have been in use for a number of years. When first introduced they, like transistors, were very expensive, and were used only for special applications. Fortunately, manufacturing techniques have advanced to a point where a single red LED costs less than $0.10. A further achievement is the availability of yellow, orange, and green LEDs.

When we think of graphics displays, we usually think of television-type video displays. All of the more popular personal computing systems have video displays, with the majority of them supporting graphics. It is not inconceivable that we will eventually see economical, flat, high-resolution LED displays which have the same capabilities as the current cathode ray tube displays. A manufacturing breakthrough will be required before this is a reality.

There have been some military programs requiring the construction of such displays. A few years ago, while still a member of the military-industrial complex, I worked on a bid to build a 10 by 10 foot LED display comprised of 792,000 discrete LEDs. My calculations at the time predicted that it would take about 3 kW of power to run.

This article is not going to describe how to replace your television screen with a flat panel LED display, but will attempt something a bit more modest. The concept of LED graphics is not that far in the future. While we're waiting for technology to catch up with interest, we can experiment with the concept on a limited scale and analyze the various logic alternatives. A side benefit is the construction of an 8 by 16 LED display as your newest peripheral device.

Light Emitting Diode Displays

We all know about LEDs, correct? They are the little red things that glow when a current is passed through them. Most of us even remember to use a resistor to limit the average current to around 20 mA. What many people don't realize is that an LED can also be driven by much higher currents if *pulsed* on and off, rather than run continuously. This is a significant fact to keep in mind when building a large LED display.

Figure 1 shows standard methods for using transistor-transistor logic (TTL) to drive LEDs. The TTL gate can be used to either sink or source current to the LED without external transistors. In general, TTL devices will sink 16 thru 20 mA, while some go as high as 50 mA. (It's best to check manufacturer specification sheets if you are unsure.) Open collector gates, shown in figures 1a and 1b, can be wired in either series or shunt configuration.

In figure 1a the circuit is completed and the LED is lit when a logic 1 is applied to the inverter input. The low-level output of the gate also provides a path to ground for the LED. Figure 1b, on the other hand, is a shunt circuit and exhibits an opposite logic. Normally current flow is through the LED, and it is lit. When a logic 1 is applied to the inverter, the resultant low output shunts the current to ground, shutting off the LED. There are advantages to both methods which I will discuss later.

Logic parts such as the 7400 NAND gate or 74LS04 inverter have active pull-up totem pole outputs. Rather than just a single NPN transistor like the open collector types, these have 2 transistors connected in series between the supply voltage V_{CC} and ground. Depending upon the logic state, only 1 of the 2 transistors will be conducting. Generally speaking, series and shunt LED drivers are more easily built with open collector devices. Figure 1d, however, cannot be accomplished with open collector logic, because this circuit depends upon the internal active pull-up resistance to source current to the LED. The exact amount of available current depends upon the logic type.

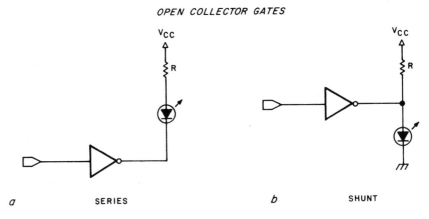

LOGIC TYPES	I_{OUT} LOW
74S $\leq$	20 mA
74H $\leq$	20 mA
74 $\leq$	16 mA
74LS $\leq$	8 mA
74L $\leq$	3.6 mA
CMOS 4049 $\geq$	3 mA
CMOS 4009 $\geq$	8 mA

OPEN COLLECTOR GATES

a SERIES

b SHUNT

ACTIVE PULLUP - TOTEM POLE GATES

EXTRA PULLUP MAY BE REQUIRED →

c SERIES

d SERIES - LED POWER SUPPLIED THROUGH GATE

Figure 1: There are several ways of driving LED displays. A method employing a series circuit with an open collector gate turns on the LED when a logic 1 is applied to the inverter input. The shunt version of the open collector circuit turns on the LED when a logic 0 is applied to the inverter input.

If active pull-up totem-pole gates are used (the kind found in nearly all TTL gates), the circuits may be wired only in series. In figure 1c the voltage needed to power the LED comes from the supply voltage V_{CC}. In figure 1d the LED is wired in series, and the power to light the LED is supplied through the logic gate. Typical output currents are given for various types of logic in the accompanying table.

Returning to the discussion of displays using LEDs, it is quite simple to take the logic concepts of figure 1 and put them to use. Figure 2 outlines a simple 8-bit LED driver with latched output. It is suitable as a bar-graph display, 8-level indicator, or 8-item annunciator. We always think first of using the video display to display the results of a logic decision, but if the result is simply yes or no, the binary answer can be signified on an LED. In my own case, such an 8-bit display is used to keep track of enabled peripherals and I/O (input/output) channels.

Larger LED Displays Have to Be Multiplexed

Using 8 LEDs probably doesn't excite too many people, especially when I started out with a number like 792,000. The 8 LEDs can, of course, be expanded to 64 by multiplying this same circuit 8-fold. With an average current of 15 mA for each LED and 100 mA for each 74100 dual 4-bit latch, the grand total to run it is slightly under 2 A at 5 V. This fact, and the necessity of having 64 resistors as well, leads us to consider some other means of driving the LEDs.

The logical alternative to continuous operation is time-multiplexed operation. For an LED with a 20 mA

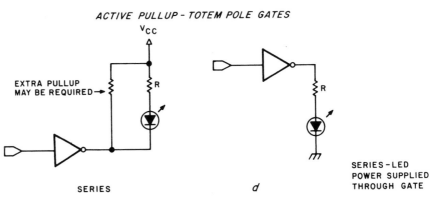

Figure 2: A simple 8-bit, latched-output LED display, suitable for use in computer-controlled bar graphs or 8-level indicators.

continuous current rating, this means we'd raise the peak current (I_{pk}) and reduce the duty cycle. If the duty cycle were 25%, then 4 LEDs could be multiplexed through the same driver, and all would appear to operate continuously. The more LEDs in the loop, the lower the duty cycle. To maintain the same brightness, the current is raised again to produce a reasonable average current. It reaches a point of diminishing returns when the duty cycle becomes so low that the peak current required to maintain a sufficient average current burns out the LED due to excessive power dissipation.

For pulsed applications, a curve of maximum peak current, pulse width, and repetition rate can be used to determine the maximum recommended operating conditions. Figure 3 illustrates a typical curve for a T-1¾ LED such as that used in this article. It is determined by comparing peak and average junction temperatures during strobed operations, and maintaining a limit equivalent to the maximum allowable DC conditions. At any specified repetition rate, the relationship between maximum current and pulse width is shown. If, for example, 5 LEDs were to be multiplexed, and brightness maintained equivalent to a 10 mA continuous current, each would have to be pulsed for 1 ms 100 times a second, with a peak current of 100 mA.

Figure 4 shows a simple 4 by 4 LED matrix which demonstrates this concept. It also serves to point out some of the limitations of this bare-bones approach. A latched 8-bit parallel output port is all that is necessary to run this display. Four bits define the column and 4 bits define the row. Multiplexing is done in software.

To turn on the LED at location A22, bits B2 and B6 would be set to a logic 1, while lighting A43 would require a combination of bits B1 and B4. The logical process is essentially an extension of the shunt circuit described in figure 1.

A microprocessor can be used to control an X,Y addressable array of LEDs. The external circuitry required is minimal, and relatively little processor time is used to refresh the array. The technique used is to periodically strobe a row and column address into an output latch. At a predetermined later time, new information concerning the next display point is sent out to the latch. If this addressing can proceed faster than 100 times per second, then the entire display will appear to be DC driven. Usually, refresh timing is handled through interrupts.

There are important considerations to keep in mind when building this type of circuit. 7406 and 7407 inverting and noninverting drivers are not high current drivers, but they can sink 40 mA. They were chosen because they are cheap and available. If brightness is a problem and peak current has to be increased, these drivers can be replaced with transistors which have a higher current rating, or more gates of the same type can be added in parallel. The fact

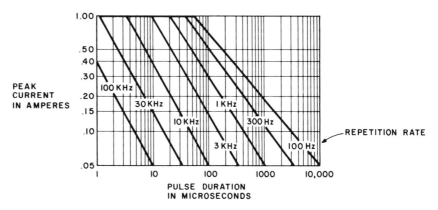

Figure 3: A typical curve for a T-1 ¾ LED showing the relationship between maximum current and pulse width for specified pulse rates.

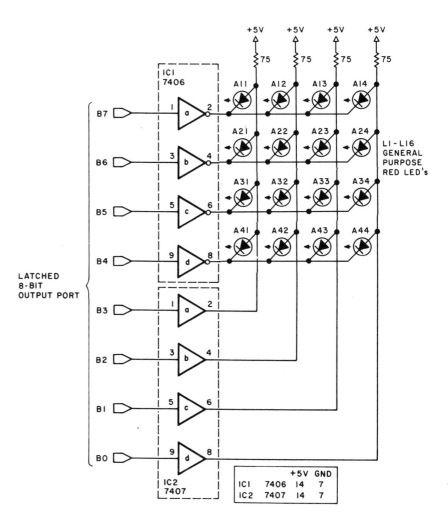

Figure 4: A simple 4 by 4 LED matrix which is software driven.

Photo 1: *The prototype board for the light emitting diode (LED) display showing all of the LEDs turned on. A piece of red plastic is held in front of the display to increase visibility.*

Photo 2: *The prototype board displaying GO→ without a red filter in front of the LEDs.*

that they are open collector devices readily allows this.

The second concern is lamp brightness. LEDs operated at low currents can have widely varying brightness. It is a good idea to pretest and select LEDs which appear to have the same intensity at a specific current.

Build a Self-Refreshing LED Display

So far I've discussed arrays which, because of their size, have limited appeal and application. A 4 by 4 display is still in the realm of indicator, rather than information display panel. To

be really effective it should at least be able to display an alphanumeric character. Such a requirement dictates a minimum matrix size of 5 by 7. This adequately displays all upper-case letters and numbers. But if you are going to have 5 by 7, why not 10 by 7 for 2 letters and so on?

At some point we have to be rational. If it were that easy to make 200 by 200 LED arrays, someone would be making them now. In my case I needed a multipurpose flat panel display that could flash a message (even if only 1 letter at a time) and serve as a sophisticated

annunciator for my alarm system. The latter was the true reason for the use of LEDs.

A transparent sheet with an outline of my security system is placed over the LED array. Significant information is indicated by flashing the LED at the point within the array that corresponds to appropriate sensor activation. It is quite interesting to watch the approach of a car down the driveway as a series of LED indicators track it.

A 4 by 4 display was too low in resolution, and while a 5 by 7 display allowed ASCII alphanumeric displays, it was also a bit limited. Considering the hardware techniques employed and relative indifference to refresh considerations, I settled on an 8 by 16 display.

Photos 1 and 2 show the completed display prototype. The prototype consists of 128 red LEDs arranged in 16 columns of 8. Photo 1 illustrates them all lit. A red plastic filter is used to enhance the display. Photo 2 shows it without the filter.

The schematic diagram for this interface is outlined in figure 5. As with the majority of my designs, I've made this to be processor and program execution-speed independent. It works equally well with assembly-language or BASIC systems, provided that a program can directly address output ports. The interface is a stand-alone peripheral. Once loaded with display data, refresh operation is locally controlled, and the computer can even be shut off without disturbing the display.

Self-Refreshing— How Does It Work?

There are 3 major hardware subsystems in the 10-chip circuit: input decoding, data storage, and refresh scanning. To the computer, this interface appears as 16 output port addresses numbered 112 thru 127 decimal (remember BASIC uses decimal

Number	Type	+5 V	GND
IC1	7430	14	7
IC2	7404	14	7
IC3	74121	14	7
IC4	7489	16	8
IC5	7489	16	8
IC6	7406	14	7
IC7	7406	14	7
IC8	74157	16	8
IC9	7493	5	10
IC10	74154	24	12

Table 1: *Power-wiring table for figure 5.*

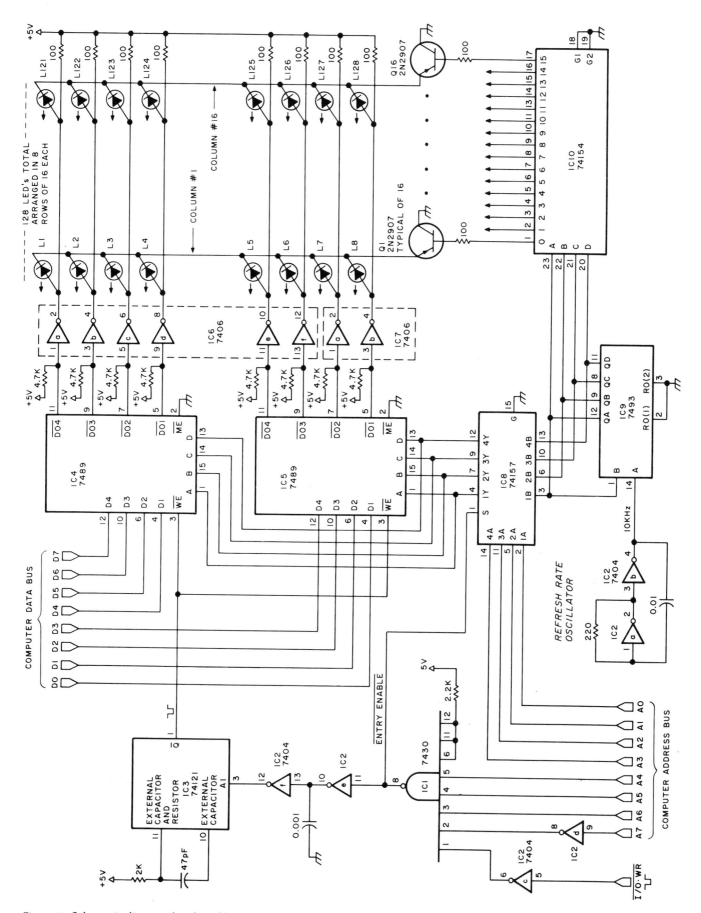

Figure 5: Schematic diagram for the self-refreshing 8 by 16 LED array display. The display is fully static and appears as 16 output ports to the computer. Note that the schematic diagram shows the use of 10 integrated circuits, while the prototype board only has 9 integrated circuits on it. The I/O decoding logic on the prototype system was not constructed on the board, but on the other end of the ribbon cable shown in the photographs.

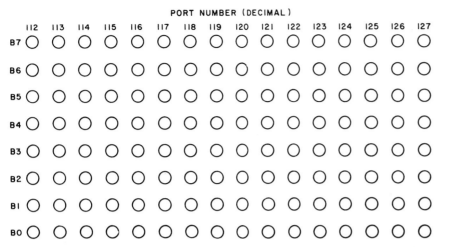

	112	113	114	115	116	117	118	119	120	121	122	123	124	125	126	127
B7	○	○	○	○	○	○	○	○	○	○	○	○	○	○	○	○
B6	○	○	○	○	○	○	○	○	○	○	○	○	○	○	○	○
B5	○	○	○	○	○	○	○	○	○	○	○	○	○	○	○	○
B4	○	○	○	○	○	○	○	○	○	○	○	○	○	○	○	○
B3	○	○	○	○	○	○	○	○	○	○	○	○	○	○	○	○
B2	○	○	○	○	○	○	○	○	○	○	○	○	○	○	○	○
BI	○	○	○	○	○	○	○	○	○	○	○	○	○	○	○	○
B0	○	○	○	○	○	○	○	○	○	○	○	○	○	○	○	○

Figure 6: The 128 light emitting diodes (LEDs) are laid out in groups of 8. Each group of 8 is assigned to a consecutive output port. The port numbers are given here in decimal.

notation). Each column represents the 8 bits of that port.

The most-significant bit (MSB) is at the top and the least-significant bit (LSB) is at the bottom. The leftmost column is decoded as port number 112 and the rightmost is port number 127. This is depicted in detail in figure 6. These selections are arbitrary and can be any 16 successive port addresses you have available. These ports can also be memory mapped to use PEEK and POKE instructions

rather than input/output instructions, if you wish. (For further information on memory mapped I/O I refer you to the book *Ciarcia's Circuit Cellar, Vol. I* from BYTE Books.) ICs 1 and 2 decode these 16 addresses.

Integrated circuits IC3, IC4, IC5 and IC8 perform the data storage function. IC4 and IC5 are each 4-bit by 16-word programmable memory devices which together form an 8-bit by 16-word storage. When data is ready for display, the computer per-

forms an output procedure to the selected port. The entry-enable line goes low, selecting address bus lines A0 thru A3 to be applied as the address inputs to the 2 memory devices.

If port decimal 115 were selected in BASIC, the binary address would be 0011. Sections c and d of IC2 are included to forestall a potential race condition and serve to delay the firing of the one-shot monostable multivibrator IC3 until the propagation delay of ICs 4, 5, and 8 is satisfied. Once this port address is set through the 74157, the one-shot fires and writes the data present on the data bus into the memory. This is essentially the same sequence as any latched output port with the exception that 16 data bytes can be stored.

The schematic diagram as shown uses transistor-transistor logic (TTL) devices. If you have an S-100 system, or otherwise have limited bus driving capabilities, you may want to substitute low power TTL devices where necessary, or buffer all incoming lines.

The final area of significance is the LED refresh scanner. Figure 7 provides an expanded illustration. Rather than successively addressing 128 LEDs, resulting in a very low-

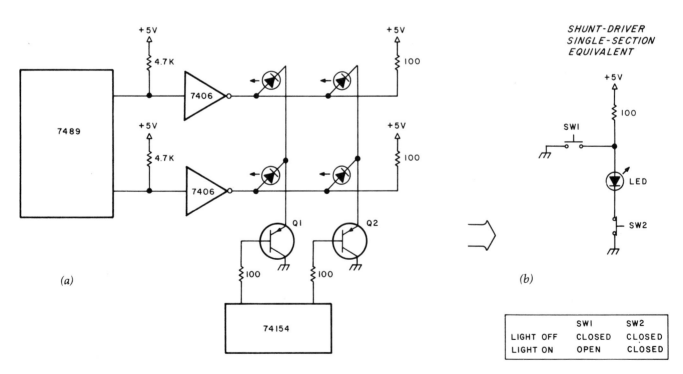

Figure 7: Expanded illustration of column scanning techniques used in self-refreshing LED graphics unit shown in figure 5. Each LED is not addressed sequentially; the LEDs are addressed by column. This results in a higher refresh rate and lower peak current to maintain a uniform brightness. For any particular LED to be turned on, the equivalent of 2 switches has to be closed (SW1 and SW2 in 7b). For this to happen in the circuit of figure 7a, the column must be addressed by the 74154, and then the coinciding byte of memory (7489) provides the other switch. The LED is lit when the correct row is addressed and the corresponding bit is set to 1.

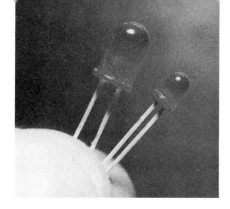

Photo 3: Two of the more popular red LEDs are the TIL-209 (the smaller LED) and the TIL-220 (larger LED). The average price for these devices is $0.09 and $0.11 apiece in quantity.

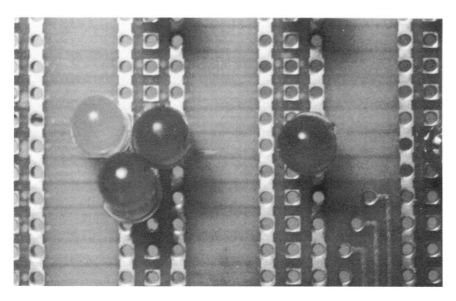

Photo 4: To experiment with 3-color displays, 3 LEDs must be placed in each position on the board.

Photo 5: The program in listing 2 produces the display shown here.

duty cycle, this design incorporates column scanning. Each light emitting diode (LED) is refreshed once every 16 clock pulses, rather than once every 128. The result is that lower peak current is required to maintain sufficient illumination.

When no data is being written into the memory (ICs 4 and 5), the address multiplexer is in the display mode. In this case it continually channels the output of a 4-bit free-running counter (IC9) to the memory address input. IC10 also receives this address and enables the particular column to which the data pertains.

In a normal sequence, the first address is 0000 binary. Since the memory is in a read condition, the output will reflect the data contents which had been stored previously as an output to port 112. IC10, a 4 to 16 demultiplexer, enables the first line by bringing it to a logic 0. The shunt drivers now enabled will allow any LED in that column to turn on in response to a stored logic 1 on that bit position. The only LEDs that can light at this time are in the first column.

The circuit will stay on this address until the next clock pulse from ICs 2a and 2b. The next address would enable the next column with similar results. The scan oscillator should be fast enough that the display does not flicker.

Various LEDs can be used. Probably the most popular size is the T-1¾ (such as the Texas Instruments TIL-220) made by most LED manufacturers and priced at about $0.11. If space is a problem, a smaller T-1 can be used with cost at about $0.09. Their relative sizes are shown in photo 3.

There is nothing which requires that the display be monochromatic. Considering that color television screens are actually discrete dots which seem to blend together when viewed from a distance, this same possibility is open for use with LEDs to a limited extent. The 3 LEDs can be mounted quite closely as demonstrated in photo 4. Experimenting with the tricolor system produced some interesting results. You must realize, of course, that a 3-color display would require 3 sets of digital logic equivalent to the circuit of figure 5.

Using a Flat Panel Display

The first thing to do after powering up and checking out the circuitry is to try to write data to it. Listing 1 is a BASIC program which sequentially exercises all 128 LEDs. Erroneous data entry can usually be traced to a too long pulse width on the one-shot (IC3).

Once the arrays have been built, you are ready for the big time— displaying a 5 by 7 dot-matrix character. Photo 5 illustrates this final achievement, and listing 2 shows the simple BASIC program required to accomplish this.

Listing 1: BASIC program to turn each light emitting diode (LED) on and off in order.

```
100 REM THIS PROGRAM CHECKS EVERY LED INDIVIDUALLY
110 REM BY OUTPUTING A SERIES OF COMPUTED VALUES TO THE
120 REM APPROPRIATE OUTPUT PORT
130 REM
140 REM 8X16 DISPLAY IS ADDRESSED AS 16 PORTS - NO.S  112 TO 127 DECIMAL
150 REM WITH LSD ON THE LEFT AND MSD ON THE RIGHT
160 REM
170 REM FIRST THE DISPLAY IS  BLANKED BY OUTPUTING ALL ZEROS
180 FOR S=112 TO 127
190 OUT S,0
200 NEXT S
210 REM
220 REM STARTING FROM THE LOWER LEFT CORNER LEDS ARE PROGRESSIVELY LIT
230 REM UP AND DOWN THE COLUMNS MOVING TOWARD THE RIGHT
240 FOR I=112 TO 127
250 FOR B=0 TO 7
260 A=2^B
270 OUT I,A
280 GOSUB 1000
290 NEXT B
300 OUT I,0
310 NEXT I
320 GOTO 240
1000 FOR T=0 TO 50
1010 NEXT T
1020 RETURN
```

Listing 2: BASIC program to write GO→ on the LED display.

```
100 REM THIS PROGRAM WRITES GO > ON THE DISPLAY
110 REM USING DATA STATEMENTS TO ENTER MATRIX DATA
120 DIM X(100):DIM S(100)
130 DATA 124,130,130,138,142,0
140 DATA 124,130,130,130,124,0
150 DATA 16,84,56,16
160 FOR S=1 TO 16
170 READ X(S)
180 NEXT S
190 FOR C=112 TO 127
200 OUT C,X(C-111)
210 NEXT C
220 STOP ;GOTO 190
```

Static displays are interesting, but if you really want to do a little crowd-pleasing, then I suggest simulating a moving marquee. Because this display interface is column-oriented, it is relatively simple to accomplish this feat. Listing 3 is a program for shifting the letter A across the display.

The character is left-justified when first displayed with the 5 by 7 data written in ports 112 thru 116. On the next programmed update, the same data is written to ports 113 thru 117, effectively shifting it to the right by 1 column. For long messages, the most effective method is to utilize a software pointer. Even a 2⅖ inch character-moving marquee is very impressive and can easily convey intelligent information.

This 8 by 16 matrix can be expanded by adding more memory and column decoders. It can be further enhanced by the addition of other colors within the same array.

The video screen need not be the only output display on a personal computer. It is only a matter of time before large arrays are commercially available, but in the meantime we can experiment with the concept. I hope that by presenting a self-refreshing interface design which eliminates the necessity of interrupts or dedicated program refresh, I may spark the interest of many experimenters. ■

Listing 3: BASIC program to move the letter A across the display from left to right.

```
100 REM THIS PROGRAM DEMONSTRATES USING THE DISPLAY PANEL AS A MOVING MARQUEE
110 REM A 5X7 DOT MATRIX LETTER A IS DISPLAYED ON THE LEFT SIDE
120 REM AND THE SHIFTED ACROSS THE DISPLAY TO THE RIGHT. USING THIS CONCEPT
130 REM VIRTUALLY ANY MESSAGE CAN BE WRITTEN.
140 DIM A(100) :DIM S(20) :DIM X(100)
150 REM FIRST THE LETTER A IS LEFT JUSTIFIED ON THE DISPLAY
160 A(1)=254 :A(2)=144 :A(3)=144 :A(4)=144 :A(5)=254 :REM A(1)-A(5) EQUAL THE LETTER A
170 FOR Q=6 TO 20 :A(Q)=0 :NEXT Q
180 REM
190 REM
200 REM CLEAR THE DISPLAY
210 FOR L=112 TO 127 :OUT L,0 :NEXT L
220 REM
230 REM
240 REM DEFINE TRANSPOSED MATRIX X(1) TO X(16) AND SHIFT RIGHT ONE COLUMN
250 S=1
260 FOR D=1 TO 16
270 X(D)=A(S)
280 S=S+1
290 IF S>20 THEN S=1
300 NEXT D
310 S=S+3
320 GOSUB 370
330 GOTO 260
340 REM
350 REM
360 REM WRITE TRANSPOSED MATRIX TO DISPLAY
370 FOR L=112 TO 127
380 OUT L,X(L-111)
390 NEXT L
400 FOR T=0 TO 300 :NEXT T :RETURN
410 RETURN
```

BUS-SIGNAL LINES

Dear Steve,

I have a Radio Shack TRS-80 microcomputer, and would like to interface your LED (light-emitting diode) display. Can you tell me what pins I should use on the TRS-80's 40-pin Expansion Interface connector?

Randy Biggs

I am glad that you want to build this device. I listed the signal names on the schematic diagram, but am happy to list the bus-signal pins as well. (See table 1.) ...**Steve**

Table 1

TRS-80	Pin Designations Signal	Apple II	S-100
20	D7	42	90
24	D6	43	40
28	D5	44	39
18	D4	45	38
26	D3	46	89
32	D2	47	88
22	D1	48	35
30	D0	49	36
36	A7	9	83
38	A6	8	82
35	A5	7	29
31	A4	6	30
34	A3	5	31
40	A2	4	81
27	A1	3	80
25	A0	2	79
39	+5 V	25	1 & 51(+8 V)
8	GND	26	100
12	STROBE	1	see figure 1

Figure 1

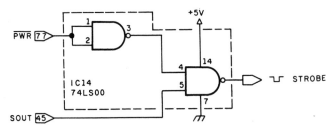

Dear Steve,

I enjoyed reading your article on light-emitting-diode (LED) graphics displays. If a display were built using optical fibers, how would the price compare with an LED-type display? Can you suggest any references? Can you suggest a circuit board (or a manufacturer) that provides high-resolution color graphics with at least a 256-by-256 pixel display?

Robert Ashworth

I am afraid, Bob, that you are trying to compare apples and bananas. Light-emitting diodes are actually light sources while optical fibers are light conductors. The latter have no self-illuminating capability. You could make my LED graphics display into a fiber-optics display. This would be done by "piping" the emitted light to a remote location using optical fibers. Since LEDs are used in both cases, the fiber optics do not make the display any cheaper.

I hesitate to recommend equipment because graphics depends heavily on the configuration of your computer system. The personal computer market is so dynamic that any suggestion I might make could be out of date by the time it was published....**Steve**

LIQUID-CRYSTAL DISPLAYS

Dear Steve,

I recently examined a Milton Bradley Microvision miniature video game, which features a 1.5-inch-square liquid-crystal display (LCD) consisting of 16 rows of 16 square blocks. I want to build a circuit to drive this display unit. How difficult would it be to modify the circuit you presented for use with an 8-by-16 array of light-emitting diodes (LEDs)? The LCD display unit could provide useful capability to a single-board microcomputer.

I have also considered developing a programmable game cartridge for the Microvision console. The console contains two 9 V battery cells, a voltage regulator, a potentiometer "paddle control," a piezoelectric beeper, a 4-by-3 printed-circuit keypad, the LCD unit, and a 40-pin dual-inline-package integrated circuit that appears to be the display driver. The Blockbuster game cartridge that comes with the console contains a 28-pin integrated circuit, a window for the display, and labeled cutouts for four control keys, along with passive components. Communication between the cartridge and the console is via a 24-pin connector.

I don't expect you to design circuits for me; if you did that for everyone who writes, you would not have enough time for your own work. However, you could do me a real favor by identifying two integrated circuits in the Microvision game. The first has 40 pins and is marked "SCUS0488, H 7920."The second has 28 pins and is marked "TMS1100NLL, MP 3450A, DBU7932."

I hope you will keep up the good work.

Daniel Q Dye Jr

A lot of people are interested in using the LCD unit you mention. However, LEDs and LCDs have very different principles of operation. An LED becomes a source of light when you pass an electric current through it, consuming a fair amount of power. LCDs, on the other hand, act as voltage-controlled reflectors of light. When an AC voltage (not DC) is applied to a liquid-crystal display, the liquid changes from transparent to opaque, consuming relatively little power. Because of this, the design approach in my LED project does not work for LCDs. But don't despair: I have written a tutorial article on LCDs.

*Concerning the components in the Microvision: the 28-pin device is a Texas Instruments TMS1000-series 4-bit microprocessor, that uses CMOS (complementary metal-oxide semiconductor) technology. The program for the Blockbuster game (or other game) is contained within it in a read-only memory. The 40-pin part is a custom multiplexed display-driver circuit for the LCD unit. The display driver is driven through the I/O (input/output) lines of the microprocessor. I hope I've helped....***Steve**

the best collection of homebrew-type construction ideas and projects available to the personal-computer experimenter. Your recent article "Self-Refreshing LED Graphics Display" has prompted me to write you.

I'd like to propose a project to you. I understand that a construction project called "Cyclops" appeared in *Popular Electronics* that actually used a dynamic-memory integrated circuit to act as a "pseudo-image sensor." Can this unique idea be extended to larger-area memory devices? The 4 K-byte circuit would make a nice 64-by-64 element array.

Jesse Newton

Thanks for the pat on the back. Sometimes late at night I need it.

I remember that article well, and I have wanted to try exactly what you suggest. I've waited because I want fairly high resolution. Perhaps with the new 32 K and 64 K bit devices I will try it. Give me a little time.

*The real problem I have is that there are so many good article ideas. I still want to put a computer in a car, do something with solar heat, remote control, and robotics. As long as you haven't been dissatisifed with everything so far, I trust that I'll find something interesting in the meantime....***Steve**

BEYOND "CYCLOPS"

Dear Steve,

I consider your series of articles

The Intel 8086

There has been a lot of talk about 16-bit microprocessors lately. You are probably interested in how they work and how they differ from present 8-bit microprocessors. This may seem more important to someone designing systems for a living rather than to the casual computer experimenter; but ultimately personal computing will be affected.

The majority of systems currently available use 8-bit processors primarily because few cost-effective 16-bit processors were available when these systems were designed. As new personal computers are conceived, the designers will have more 16-bit microprocessors to choose from, and in my opinion, the latter will win out.

Software development is much more expensive than hardware development. It is much cheaper to write one line of code executing a hardware multiply instruction than to write an algorithm to do the same function on a processor devoid of this direct capability. Reduced cost of development should be reflected in lower retail cost. There are always exceptions to the rule, but once amortized and in volume production, the 16-bit microprocessor should prove to be the logical choice for medium- to high-level applications.

The Intel 8086

It isn't necessary to wait any longer if you have a burning desire to learn about 16-bit microprocessors. The latest one available and in volume production is the Intel 8086. The 8086 is a 16-bit microprocessor which is upward-compatible from the 8-bit 8080/8085 series processors. The 8086 contains a set of powerful, new 16-bit instructions. This enables a system designer familiar with 8080 devices to start coding immediately and gradually gain expertise in using the additional 16-bit instructions. It is important to realize that when I refer to compatible instructions I mean functional compatibility. A program written for an 8080 would have different object code than an 8086. This is only a slight inconvenience considering that this former 8080 program should run about ten times faster on an 8086. The evolutionary step between the 8086 and 8080 is far greater than that between the 8080 and 8008.

The apparent goal of Intel designers was to extend existing 8080 features symmetrically and add a wide range of new processing capabilities. The added features include 16-bit multiply and divide, interruptible byte-string operations, 1 M byte direct addressing, and enhanced

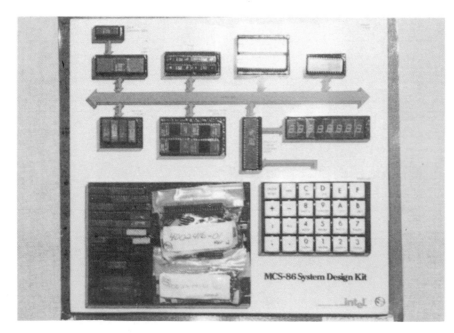

Photo 1: *SDK-86 system as delivered from factory.*

bit manipulation. Arithmetic operations are accomplished in American Standard Code for Information Interchange (ASCII) or binary-coded decimal with a one-instruction hardware conversion.

In addition to the capability of handling data in bits, bytes, words, or blocks, the 8086 incorporates many features formerly found only in minicomputer architecture. It also supports such operations as reentrant code, position-independent code, and dynamically relocatable programs.

The 8086 is fabricated with a newly developed, high-speed metal-oxide semiconductor (H-MOS) process which is considerably faster than standard MOS. Running up to 8 MHz, the 29,000-transistor 8086 is the fastest single-chip central processor currently available. Unlike the 8080/8085 processor's registers, the 8086's registers can process 16-bit as well as 8-bit data.

Figure 1a shows an internal block diagram of the 8086. The 16-bit arithmetic/logic instructions are handled within the general register files. This section contains four 16-bit general data registers, two 16-bit base pointer registers, and two 16-bit index registers. Figure 1b illustrates an 8086 register model for comparison to the 8080.

The four data registers, addressable also in 8-bit partitions, are primarily from the original 8080. There are twice as many general-purpose registers as there are on 8-bit processors.

The relocation register file is the other unique 8086 enhancement. This group is referred to as the segment register file, and extends direct addressing capability to a full megabyte of memory. This file has four address pointers which contain program relocation values for up to four 64 K byte program segments. In addition, a fifth pointer serves as an I/O (in-

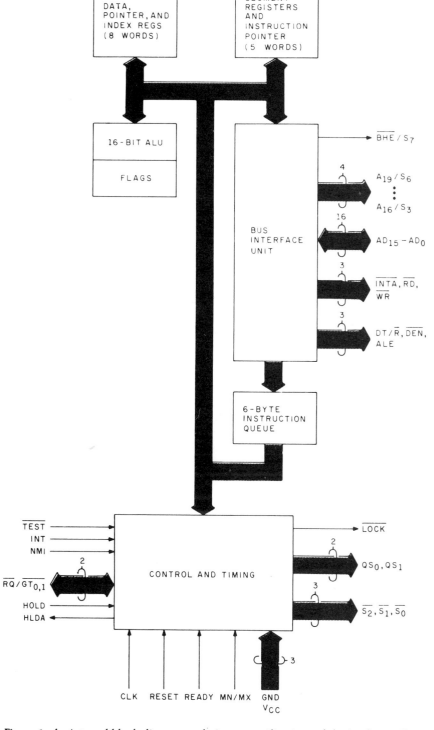

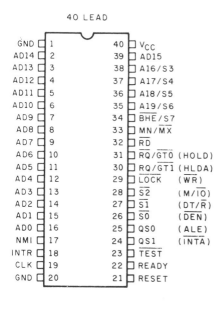

Figure 1: *An internal block diagram and pinout specifications of the Intel 8086 (figure 1a). Figure 1b shows the 8086 register model illustrating the differences between the 8086 and the 8080. Figure courtesy Intel Corp.*

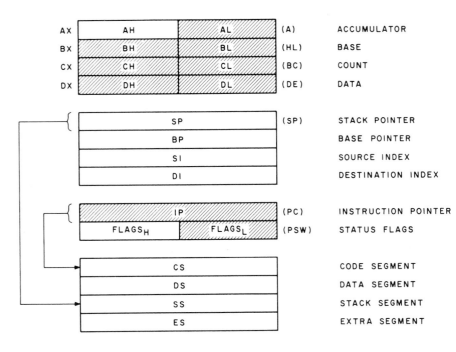

put/output) control providing address space for a full 65,536 I/O ports.

Logically the 8086 operates more like larger computers than like a classical microprocessor. This is accomplished through independently controlled bus interface and execution units (figure 2). The major contribution is to speed processing by overlapping instruction fetch and execution. Up to six bytes of instruction are placed in a queue before execution. As each instruction is processed, the following instructions move up one position and a new instruction is fetched and placed in the queue. This simultaneous fetch and execute capability induces more efficient use of the memory bus. It is possible for two single-byte 8086 instructions to be executed within the time for one memory cycle. The result is improved performance, given the same bus bandwidth and memory speed as other systems.

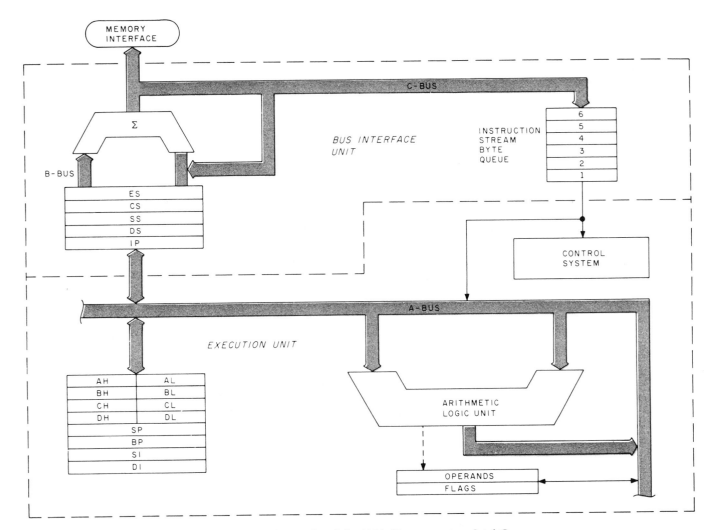

Figure 2: *Functional block diagram of internal data paths of the 8086. Figure courtesy Intel Corp.*

The Intel SDK-86

Perhaps this brief introduction has sparked your curiosity and you wish to know more about the 8086. Of course, the best method of learning is to use one. Since at this writing the 8086 is still so new that it is not incorporated into any general-use personal computer, we are left to our own resources and construction abilities. Fortunately Intel realizes that the success of any new product depends on evaluation by as many potential users as possible. For this reason the System Design Kit (SDK) series of products were conceived.

The SDK-86, shown prior to assembly in photo 1, is a single-board, 8086-based computer. Intel's pricing policies make the purchase of the SDK-86 kit far more attractive than a single 8086 chip. It results, in the name of advertising, in one of the better computer offerings on the market. At $780 the SDK-86 fits within most budgets. It is a complete computer including processor, programmable memory, read-only memory, I/O (input/output), and display. Table 1 is a more explicit listing of specifications and figure 3 is a detailed block diagram.

The SDK-86 is very easy to assemble. As shown in photo 2, it comes packaged so that all components are easily recognizable, even for a novice. Documentation includes an Assembly Manual, User's Manual, User's Guide, and Monitor listings (see photo 3). The assembly procedures are written at such a level that even a person having limited technical knowledge can assemble the kit. The assembly manual progresses from basic solder techniques and component identification to step-by-step assembly and checkout. The only microcomputer assembly literature I have read which was as easily understandable as this comes from the Heathkit people.

All major components are socketed, but to be on the safe side it is a wise idea to purchase additional integrated-circuit sockets. This will allow all integrated circuits to be removed in case troubleshooting is necessary. The fully constructed com-

Photo 2: *Typical page from the construction manual. Each instruction step is clearly explained and each component is accurately identified.*

Table 1: *Summary of specifications for the SDK-86 board.*

Central Processor

Processor: 8086
Clock Frequency: 2.5 MHz or 5 MHz (jumper selectable)
Instruction Cycle Time: 800 ns (5 MHz)

Memory Type

Read-Only Memory: 8 K bytes
Programmable Memory: 2 K bytes (expandable to 4 K bytes)
(2 bytes equal one 16-bit word)

Memory Addressing

Read-Only Memory: FE000 thru FFFFF
Programmable Memory: 0 thru 7FF (0-FFF with 4 K bytes)

Input/Output (I/O)

Parallel: 48 lines (two 8255As)
Serial: RS232 or current loop (8251A)
Data Transfer: Rate selectable from 110 to 4800 bps
Display: On-board, 8-digit, light-emitting diode (LED) readout

Interface Signals

Processor Bus: All signals transistor-transistor logic (TTL) compatible
Parallel I/O: All signals TTL compatible
Serial I/O: 20 mA current loop or RS232

Interrupts

External: Maskable and nonmaskable; Interrupt vector 2 reserved for nonmaskable interrupt (NMI)
Internal: Interrupt vectors 1 (single-step) and 3 (breakpoint) reserved by monitor

Direct Memory Access

Hold Request: Jumper selectable, TTL compatible input

Software

System Monitors: Preprogrammed 2316 or 2716 read-only memories
Addresses: FE000 thru FFFFF
Monitor I/O: Keypad and Serial (teletypewriter or video display)

Power Requirements

V_{cc}: + 5 V (± 5%), 3.5 A
V_{TTY}: − 12 V (± 10%), 0.3 A (required if teletypewriter (TTY) or video display terminal connected to serial interface port)

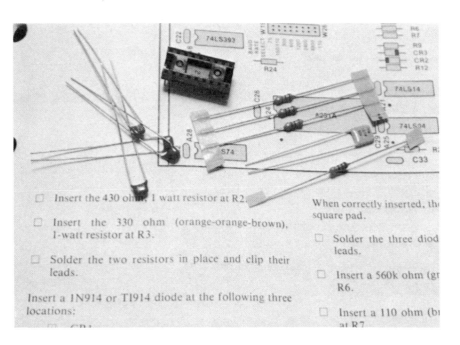

☐ Insert the 430 ohm, 1 watt resistor at R2.

☐ Insert the 330 ohm (orange-orange-brown), 1-watt resistor at R3.

☐ Solder the two resistors in place and clip their leads.

Insert a 1N914 or T1914 diode at the following three locations:

When correctly inserted, the square pad.

☐ Solder the three diode leads.

☐ Insert a 560k ohm (gr R6.

☐ Insert a 110 ohm (br at R7.

puter is shown in photo 4. Checkout, after determining that there are no obvious errors, is simply a matter of applying power and pressing the system reset button.

When the SDK-86 is reset, the 8086 executes the instruction at hexadecimal location FFFF0. The instruction at this location is an intersegment direct jump to the beginning of the monitor program that resides in read-only memory, hexadecimal locations FF000 to FFFFF. The monitor is comprised of two programs resident in programmable read-only memory; one for use with the on-board keypad, and the other a serial monitor that supports a video display or teletypewriter connected to the Electronics Industries Association (EIA) serial interface connector. This latter communication mode is preferable if the SDK-86 is to be used efficiently for software development. Even though the system is constructed to vector to the keyboard monitor on power up, simply interchanging the two sets of programmable read-only memory will allow the unit to start up immediately in the serial mode.

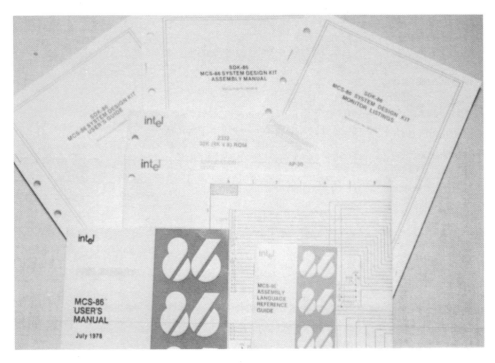

Photo 3: *The SDK-86 board comes complete with well-written documentation manuals for assembly and use.*

The SDK-86 Monitor

Both monitors share similar command capability. The keyboard monitor is optimized for the 8-digit, light-emitting-diode (LED) display while the serial monitor is obviously for a video display or teletypewriter. The only dissimilarity is that the latter has the additional ability to read or write to a paper-tape punch, or with the addition of a Frequency-Shift-Keying (FSK) modulator/demodulator, cassette storage. Table 2 lists the serial monitor I/O commands.

Of particular importance are the single-step and go commands. Single step allows a program to be executed one instruction at a time, while the go command allows the user to specify a breakpoint which returns control to the monitor while preserving the machine's status. This allows a program to be run in segments facilitating checkout.

While the monitor does provide some powerful routines, the PL/M listings provided in the documentation do not directly give the addresses of the individual routines. Enough effort is required to extract this information, that rewriting particular routines in user memory is a worthwhile consideration.

In Conclusion

If you have an interest in 16-bit microprocessors, perhaps the best

Photo 4: *Assembled SDK-86 board. Note the prototyping area on the left-hand side.*

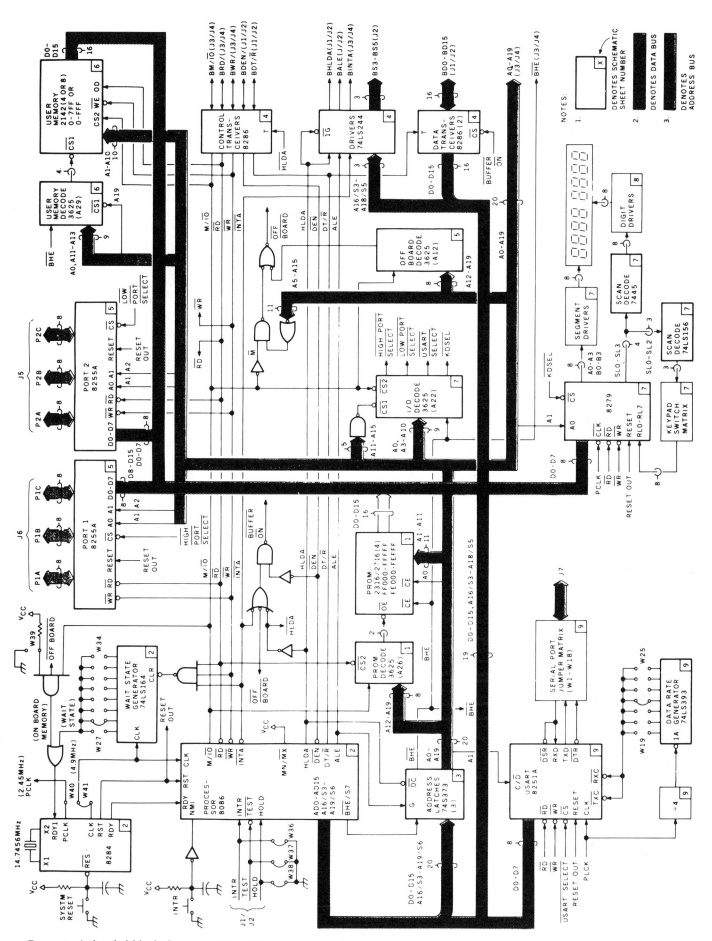

Figure 3: *A detailed block diagram of the SDK-86 evaluation board. Figure courtesy Intel Corp.*

Table 2: *The commands which are available for use with the serial monitor.*

Command	Monitor Command Summary FUNCTION/SYNTAX
S (Substitute Memory)	Displays/modifies memory locations S[W]<addr>,[[<new contents>],]*<cr>
X (Examine/Modify Register)	Displays/modifies 8086 registers X[<reg>][[<new contents>],]*<cr>
D (Display Memory)	Moves block of memory data D[W]<start addr>[,<end addr>]<cr>
M (Move)	Moves block of memory data M<start addr>,<end addr>,<destination addr><cr>
I (Port Input)	Accepts and displays data at input port I[W]<port addr>,[,]*<cr>
O (Port Output)	Outputs data to output port O[W]<port addr>,<data>[,<data>]*<cr>
G (Go)	Transfers 8086 control from monitor to user program G[<start addr>][,<breakpoint addr>]<cr>
N (Single Step)	Executes single user program instruction N[<start addr>],[[<start addr>],]*<cr>
R (Read Hexadecimal File)	Reads hexadecimal object file from tape into memory R[<bias number>]<cr>
W (Write Hexadecimal File)	Outputs block of memory data to paper tape punch W[X]<start addr>,<end addr>[,<exec addr>]<cr>

place to start is with the SDK-86. The 8086 is a quantum leap forward for microprocessors and the SDK-86 is a cost-effective method of evaluation, complete with all the hardware of a basic computer system. It must be cautioned that a first-time user, unaccustomed even to 8-bit microprocessors, may find the learning process somewhat complicated. The SDK-86, while packaged and assembled in a Heathkit fashion, is an industrial training device and not aimed specifically at the personal computing market. Beyond the minimal checkout procedures and brief description of the monitor commands, there are no sample programs which can be immediately entered and executed. This unit must be thought of as a rather sophisticated trainer. The mechanism is provided in the form of the board, but the actual course of education is completely in the hands of the user. ∎

SDK-86 INQUIRIES

Dear Steve,

I am a subscriber to *BYTE*, and I have enjoyed reading your articles for over two years. Your articles have increased my knowledge of digital circuitry and microcomputers. Thus, one purpose of this letter is to thank you for your effort. Although I constantly read articles in *BYTE* and other technical magazines, I am only now thinking of assembling my own computer. Perhaps you could answer some of my questions:

In your article on the Intel SDK-86 computer kit, the data-rate generator is fed by a 612,500 Hz clock. It seems to me that the 8-bit counter (a 74LS393) would divide this by 256 to produce a minimum rate of over 2 kHz. Where does the 110 bps (bit per second) rate come from?

I am considering the purchase of an Intel SDK-85 kit and a Heathkit H-19 (smart video terminal). I believe that they will be compatible; how hard can the interfacing be? Since the serial I/O (input/output) port of the SDK-85 runs at 110 bps, it seems that the initial loading of the H-19 may take as long as 3 minutes. What is the best way to interface a printer to the computer at the same time?

I am interested in obtaining BASIC firmware; I have seen advertisements for BASIC stored in ROM (read-only memory), but it seems that it may be written for a specific computer system, rather than the 8085 microprocessor in general. Can I get firmware compatible with the SDK-85 computer that will handle I/O? Is the performance increase of the SDK-86 over the SDK-85 really worth $550?

Chin Y Chang

Thank you for the vote of confidence. I'll try to answer your questions in order:

On the SDK-86 computer, the data-rate generator is fed by a 1.8432 MHz clock. The 74LS393 and other circuitry reduce this to approximately 1760 Hz (actually a bit higher) to provide 110 bps. This unit can go as high as 4800 bps, with the change of a few jumpers.

The H-19 and SDK-85 could communicate serially. Provision is made on the SDK-85 board for the addition of an MC1488 and an MC1489 (quad line driver and quad line receiver, respectively) for RS-232 operation. Since the only data rate is 110 bps, things will indeed be slow, unless you write your own I/O routines. Interfacing to a printer requires knowledge of the printer's specifications. If it communicates serially, then a switch would allow you to use the printer in place of the video monitor quite easily. Selection of the best printer for interfacing is dependent upon your programming abilities.

Lawrence Livermore BASIC is available in read-only memory from a few manufacturers (such as National Semiconductor). Call National's local sales offices for details. The memory devices contain only the BASIC interpreter, but no I/O routines; compatibility with the SDK-85 system will, again, depend on your abilities.

*The SDK-86 is not aimed at the experimenter market. While you may benefit in the long run, your questions suggest that you might be biting off a little too much. If you want a 16-bit computer, save the $1000 of an SDK-86 kit and put it towards an assembled system....***Steve**

Add Nonvolatile Memory to Your Computer

"You know, Ray, sometimes I think I see more of you than your wife does."

He grinned and retorted, "I just dropped over to see what the Circuit Cellar Frankenstein was cooking up this month."

I fully deserved that. Few have seen the Circuit Cellar, and it does look a little imposing at first. The usual 20-square-foot hobby corner used by most computerists has been expanded to a 1000-square-foot computer room which vaguely resembles the bridge of the starship *Enterprise*. Accented with the eerie appearance of seven video displays and a multitude of strange black boxes emanating menacing sounds, it sometimes becomes an environment of computerized insanity. While I am not interested in creating any monsters, Baron von Frankenstein and I may have a few interests in common. His demise, I assure you, was simply a case of bad press.

"Steve?" Ray said loudly. "What are you working on?"

I was jerked back to reality somewhat abruptly. Visions of a 1932 movie set faded as I turned in my swivel chair to respond.

"I'm actually working on several ideas, Ray, but the easiest is trying to put a computer in a car."

Ray quickly cast a doubtful glance at the 64 K-byte, dual-disk, Z80-based system. Returning his attention to me he quipped, "Where do you plan to put the printer?"

"I don't mean a big computer. I mean a little one, probably a single board. I will have sensors throughout the car to monitor engine speed, temperatures, pressures, and so forth fed to a display visible to the driver. The driver will be able to calculate and keep running totals of gas mileage, monitor the engine performance, and generally maintain a comfortable feeling of safe motoring."

Ray said, "That sounds pretty good. You will obviously have to use CMOS for your computer." His observation was based on his long years of technical experience.

"Why?"

Ray seemed confused at my reply. He expected agreement. Shouldn't complementary-metal-oxide semiconductor (CMOS) components be used since the computer will be battery powered?

"Because it is battery powered. That's why!" he demanded.

"That is not necessarily true." Trying not to seem quarrelsome, I continued. "Let's think about an automobile for a minute. There are many power-consuming devices. A defroster fan or rear-window heater can draw 100 watts each. Without the floppy-disk drives, even that big Z80 system over there doesn't pull that much. I'm shooting to stay under 20 watts, but logic type doesn't make much difference."

"Yes, I know when the engine is running there is plenty of power available from the alternator." Ray seemed a bit frustrated in pointing out my misjudgement of the facts. He persisted. "At 12 volts, 20 watts is almost 2 amperes! The 12-volt car battery won't last long with the engine off."

"You do not leave the defroster fan on with the engine off, do you?" I countered.

"Of course not! But what about your program? If it is written into programmable memory such as 2102 or 2114 devices, you'll lose it when the power goes off."

"I said this computer is for automotive use. It can't be considered as a general-purpose computer. Rather than using only programmable memory with programs loaded from tape or disk, it will have the operating system and language interpreter stored in read-only memory and application software stored in erasable, programmable read-only memory, an EPROM. The only programmable memory needed will be a scratch pad for calculations and a

ELECTRICALLY ALTERABLE READ ONLY MEMORIES

FUNCTION	DESCRIPTION	PART NUMBER	PAGE NO.
82 BIT EAROM	82 bits org... d 82 × 1	ER0082	4-5
512 BIT EAROM	512 bits ...	ER2051	4-16
		ER2051 HR	4-16
	512 bits org...	ER2055	4-19
1K RAM/EAROM	1,0... its organize...	ER1711	4-11
1400 BIT SERIAL EAROM	1... d 100 × 14	ER1400	4-8
4K EAROM	4,096 b... × 4	ER2401	4-22
		ER2402	4-22
		ER3400	4-34
8K EAROM	8,192 bits organized 2,048 × 4	ER2805	4-28
		ER2810	4-28

Photo 1: *Shown here are the General Instrument ER3400 and ER1711 electrically alterable read-only-memory (EAROM) parts. The ER3400 has only the EAROM function; the ER1711 combines the functions of programmable memory and EAROM in a single part.*

small area set aside for storage of continually updated, long-term quantities like mileage and gas consumption."

Ray was not incorrect. The language interpreter and application program *could* reside entirely in programmable memory as in most older personal computers. Memories composed of 2102A devices can, in fact, be put in a stand-by mode by dropping the power-supply voltage to 2 V. This reduces power consumption by 80% while still retaining data.

While being practical under most circumstances, stand-by operation is not a total solution. Depending upon the system configuration, memory power might have to be sequenced on and off, and perhaps be isolated from the 5-V supply of the rest of the system.

The ultimate value of this technique is dependent on efficient power conversion. A 12-V-to-2-V converter with 50% efficiency would be self-defeating. Battery power, however much reduced, is still required as long as memory data is to be retained.

Hybrid computers exist wherein high-speed, bipolar microprocessors are mated with low-power CMOS memory. Such a system could have various forms of read-only memory as before, but in the form of complementary-metal-oxide devices rather than bipolar memory chips.

Data retention time would probably be an order of magnitude longer, but there are complications associated with mixing logic families. All things considered, if it were not for the high cost of CMOS memory, I would use it.

Ray persisted, saying, "You still have to provide continuous power for *some* programmable memory if you intend to store those long-term values." Ray was convinced that I must have constant power on something. I hated to disillusion him.

I explained, "Not necessarily. I considered the usual standby-mode programmable memory, both CMOS and bipolar types, and rejected them. Bipolar standby takes too much power; and CMOS memory chips might be destroyed by physical handling. This is experimental, you know."

Ray said, "If you intend to shut off the computer entirely, then I suppose you could write out data to an audio cassette to be reloaded when you start the car."

I said, "Well no, I want this to be automatic. I should be able to get in the car, turn the key and have the computer start too. The tape player in the car is for Bartok, not for Kansas City data." I paused slightly to allow the air to clear, and then continued. "I've been thinking of using nonvolatile programmable memory."

Ray's jaw dropped. "Nonvolatile? Do you mean *core* memory?"

While magnetic core memory is indeed nonvolatile, that was not what I meant. "No, not core, but *semiconductor, nonvolatile programmable memory!*"

Though Ray is quite technically aware, this was a new concept for him and he wasn't sure if I was serious. I have been known to play jokes like this before.

To convince him that I was serious, I began to explain, "Specifically, I am talking about *electrically alterable read-only memory,* or EAROM. You should consider it as a *read-mostly* memory."

"Well, that's different!" Ray exclaimed with relief. "You didn't say read mostly!"

All About the EAROM

EAROMs are *word-alterable* read-only memories intended for use as "read-mostly" memories. On the surface this may sound similar to an EPROM (erasable, programmable read-only memory). Once erased under ultraviolet light, an EPROM is indeed a word-alterable read-only memory.

In reality, there is very little similarity between the electrically alterable and erasable, programmable memory. An EPROM can be erased only in block mode and generally takes about 10 minutes to erase. While some can be programmed in as little as 50 seconds, an 11-minute (or more) read/write cycle-time hardly qualifies it in the category of high-speed programmable memory. An EPROM, therefore, is just a conveniently reprogrammable read-only memory.

An EAROM, on the other hand, does not rely upon ultraviolet light exposure for erasure. Clearing memory for reprogramming is done electrically. With a read time equivalent to a high-speed EPROM, complete or partial erasure in 10 ms, and a write time of a mere 1 ms, an EAROM fills the gap between truly programmable memory and EPROM. EAROM can be integrated into the memory address space of practically any microcomputer. Like regular read-only memory, it retains data (for up to 10 years) when the power is removed, and is a natural choice for bootstrap program-storage applications. Should the stored program have to be changed, you can erase the chip with a 10-ms eraser routine and then rewrite the data at a rate of 1 ms per byte. This can all be done without removing the part from the system.

There are many EAROMs available, but like other types of memory devices, their architectures and capacities vary. You would not use 1 K-bit memory chips if you had to fit 64 K bytes of memory on a small printed-circuit board, nor would you choose to use an erasable programmable read-only memory requiring a 3-voltage power supply if only a 5-V supply is available.

It is important to observe that EAROMs also have limitations. Unlike regular programmable memory, the electrically alterable read-only memory cannot be erased and reprogrammed without limit. The General Instrument model ER3400 EAROM, for instance, can have each byte *read* 2×10^{11} times, but written only 100,000 times. If this EAROM were being used as standard programmable memory in a frequently executed loop, 100,000 erase and write cycles would take only 20 minutes.

A0 thru A9	10-bit word address.
D0 thru D3	Data input and output pins.
$\overline{CE}$	Chip enable. Chip selected when $\overline{CE}$ is pulsed to logic 0.
C0,C1	Mode-control inputs.
$\overline{WE}$	Write enable. Input data read when $\overline{WE}$ is pulsed to logic 0.
V_{ss}	Substrate supply. Normally at +5 V.
V_{GI}	Ground input.
V_{DD}	Power-supply input. Normally at -12 V.
V_{GG}	Power-supply input. Normally at -30 V.

Table 1: *Functions of pins on the General Instrument ER3400 electrically alterable read-only-memory device.*

The ER3400 is better used to store tables and calculated results that must be retained if the power fails. The specific, useful qualities of the ER3400 are high density of data storage, high speed, and long time of data retention.

A Hybrid Memory Device

Where nonvolatile memory is required to have more frequent write cycles, the General Instrument ER1711 should be used. This device combines two types of memory on a single chip: a standard 1 K-bit, static programmable memory and a 1 K-bit electrically alterable read-only memory.

The programmable memory is mapped as standard program memory. There is no limitation on write cycles. When the memory contents are to be retained, such as at the time of system power-down, a sensing circuit pulses the EAROM write line. With one pulse the entire content of the 1 K programmable memory is written in parallel to the EAROM. The EAROM section has the same write-cycle limitations as the ER3400 part.

Devices such as the ER1711 are particularly suited for storing frequently changing data. As long as power is available to the system, this data resides in the programmable memory. Only during periods when the power is off or during special events is the data transferred to nonvolatile storage.

Design Choices

Each of the two EAROM devices has its good and bad features. The designer choosing EAROM parts for a memory system will have to choose among the various positive attributes and complications. No single EAROM part necessarily fits all applications. The application must determine the choice.

Failure to understand this fact and

Figure 1: *Pinout designations of the twenty-two pins on the General Instrument ER3400 electrically alterable read-only memory (1a), and a block diagram of the major circuit sections within the part (1b).*

(1a)

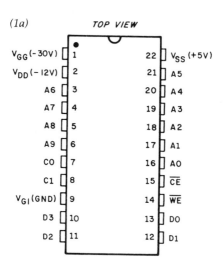

(1b)

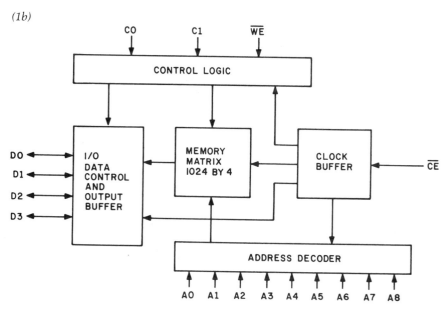

lack of adequate knowledge of the variety of parts available are contributing factors to the absence of electrically alterable read-only memories in personal computers. The two devices I have chosen to discuss should cover most applications, but other configurations do exist.

General Instrument ER3400

The ER3400 is a 4 K-bit EAROM configured as 1024 by 4 bits. 1 K bytes of nonvolatile storage can be obtained using two chips. Figure 1a shows the pinout designations, and figure 1b shows the block diagram of the chip. Photo 1 shows the part in its dual-in-line package. Table 1 describes the functions of the pins found on the ER3400.

The ER3400 requires three power-supply voltages (+5 V, −12 V, and −30 V) for complete operation.

Because of the relatively low cur-rents required to write data into an EAROM, the −30-V supply is usual-ly derived from either the +5-V or −12-V supply. A previous Circuit Cellar article ("No Power for Your Interfaces? Build a 5 W DC-to-DC Converter," October 1978 BYTE, page 22, or page 1 in my book *Ciarcia's Circuit Cellar, Vol. I*, BYTE Books, 1979) covered both theory and design procedures, should you need to make −30-V supply.

Unlike a regular programmable memory which has only read and write functions, the ER3400 has four operational modes: read; write; word erase; and block erase.

Operational Modes of the ER3400

Erase: To erase one word, both of the C0 and C1 mode-control input lines are set in the logic 1 (high) state, and the desired address location is set. A negative excursion of the

voltage on the chip enable ($\overline{CE}$) line loads in the address and control, and initiates the erase operation. To avoid tying up the microprocessor bus, this mode is latched on the positive-going edge of the $\overline{CE}$ signal. The erase operation will continue while $\overline{CE}$ is high.

When it is desired to erase the entire device, the operation is the same, except that the C0 mode-control input is low while C1 is kept in a high logic state.

A "dummy read" operation is re-

Number	Type	+5 V	GND	-12 V	-30 V
IC1	HM-7603	16	8		
IC2	ER3400	22	9	2	1
IC3	ER3400	22	9	2	1
IC4	74121	14	7		
IC5	7400	14	7		

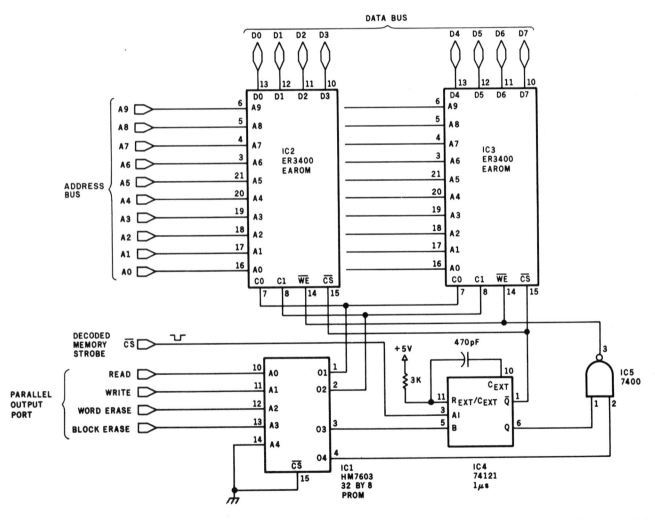

Figure 2: *Schematic diagram of a circuit that uses two ER3400s to form a 1024-byte memory. Read, write, and erase functions of the ER3400 are available using this circuit. The programmable read-only memory, IC1—the Harris Semiconductor HM-7603, a 32-word-by-8-bit PROM, serves to decode mode-control inputs received from a parallel output port. The truth table for this PROM is shown in table 3.*

quired to end the erase cycle.

Write: The control code for write is for the C0 line to be high while C1 is low. The control word and address are strobed in at the occurrence of the $\overline{CE}$ (chip enable) pulse. Data is strobed in during the write enable ($\overline{WE}$) signal. The timing requirements for the write enable signal are designed so that $\overline{WE}$ may be generated by combining the chip enable signal and a write signal through a logic gate.

As is the case with the erase operation, the control code and address are latched on the rising (positive-going) edge of $\overline{CE}$. Data is latched by the rising edge of $\overline{WE}$. As in erase, a dummy read is required to end the write cycle.

Read: To read out data, C0 and C1 control lines are both held low and the desired address is selected. The chip enable signal strobes in the mode and address data, and clocks out the data.

In all modes, when $\overline{CE}$ is high, the

C0	C1	Mode	Explanation
0	1	Block erase	Erase operation performed on all words.
1	1	Word erase	Stored data is erased at addressed location.
0	0	Read	Addresses data read after leading edge of $\overline{CE}$ pulse.
1	0	Write	Input data written at addressed location.

Table 2: *Selection of modes of operation of the ER3400 EAROM. The indicated logic levels are presented to the two mode-control inputs C0 and C1 to produce the corresponding mode of operation of the memory device.*

	Address					Data							
Operation	A0	A1	A2	A3	A4	O1	O2	O3	O4	O5	O6	O7	O8
Read	1	0	0	0	0	0	0	1	0	X	X	X	X
Write	0	1	0	0	0	1	0	1	1	X	X	X	X
Word Erase	0	0	1	0	0	1	1	1	0	X	X	X	X
Block Erase	0	0	0	1	0	0	1	1	0	X	X	X	X
All other locations						0	0	0	0	X	X	X	X

Table 3: *Truth table for the mode-decoding, programmable read-only memory (PROM) that appears as IC1 in figure 2. A high logic state is represented by 1; a low logic state is represented by 0. Where the logic state does not matter, X characters appear in the table. Possible substitutes for the Harris Semiconductor HM-7603 PROM used here include the 74S288, 82S123, and AM27S09.*

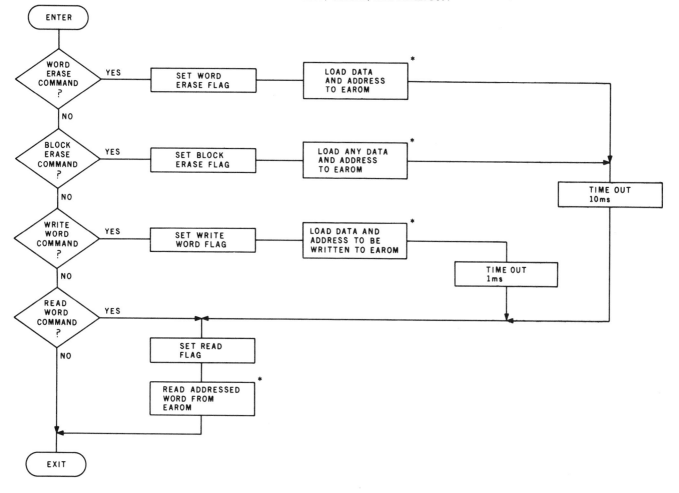

Figure 3: *A flowchart of the algorithm for actuating the various operating modes of the ER3400 interface circuit of figure 2. The diagram blocks marked with asterisks refer to standard memory read or write instructions.*

Table 4: *Descriptions of pin functions of the General Instrument ER1711 hybrid EAROM/programmable memory part.*

Pin	Description
D0 thru D3	4-bit data word
A0 thru A7	8-bit data address
V_{SS}	+5-V supply voltage
V_{DD}	−12-V supply voltage
V_{GI}	Ground
CE	Chip enable: strobed during programmable-memory cycles, held low during EAROM cycles
H	Hold: normally high, low during data-recall cycle
W	Write: high during programmable-memory write cycle
RS	Restore: normally low, pulsed high during recall cycle
E/W	Erase/Write: controls EAROM memory cells.

Figure 4: *Pinout designations of the General Instrument ER1711 hybrid electrically alterable read-only-memory/programmable-memory device.*

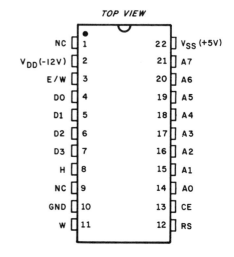

data input/output lines are in a high-impedance state. The control line logic levels for the several modes are summarized in table 2.

In the write and erase (both word and bulk) modes, the data, addresses, and the state of the control lines are loaded into internal registers within the ER3400 on the rising edge of the CE pulse, and are later cleared by a dummy read pulse also strobed by CE.

Observe carefully the requirement for the dummy read operation. Since there is no internal timing or sequencing, the ER3400 relies upon the user to terminate the write and erase functions by switching the EAROM into the read mode after 1 ms following a write and after 10 ms following an erase. There are various software and hardware methods to accomplish this.

For personal computer use, the circuit of figure 2 should be considered. A parallel output port provides the flag input to the mode-decoding, programmable read-only memory (PROM) IC1. A memory map of the mode-decoding memory is shown in table 3.

With the electrically alterable read-only-memory device attached to the address and data lines, a read operation is accomplished by setting the output bit corresponding to the A0 input line of the PROM to a high logic state. Each of the four address lines A0 thru A3 is called a flag for purposes of explanation.

Setting the read flag (A0) corresponds to a binary 10000 input code to the PROM (IC1 in figure 2). The O1 and O2 outputs will place low logic levels on both mode-control inputs C0 and C1. This places the EAROM in the read mode.

Next, the EAROM is addressed through the normal CS line. This action fires the one-shot (monostable multivibrator, IC4 in figure 2) which clocks the data onto the computer data bus. The ER3400 has a cycle time of 1.8 µs. Depending upon the memory speed of your computer, it may be necessary to add wait states when addressing EAROM.

The sequence is similar for write and erase operations. By setting the write flag (A1), the mode-decoding PROM causes C0 to go high while C1 stays low, placing the EAROM in the write mode. Addressing the appropriate byte and strobing the CS line causes the ER3400 to enter a write condition. (The byte addressed for writing should have previously been erased through either the word or block erase operations.)

After 1 ms, the read flag (A0) is set and a read sequence is executed to stop the write activity. A flowchart of the mode-selection algorithm is shown in figure 3.

Operational Modes of the ER1711

The ER1711 operates quite differently from the ER3400. Configured as 256 by 4 bits, this device combines the properties of both regular programmable memory and electrically alterable read-only memory on a single chip. Figure 4 shows the pinouts of the ER1711, and table 4 describes the pin functions.

In general, EAROM writing is rather slow. Read access time can be less than 1 ms, but writing and erasing take 1 to 50 ms (depending upon the device). As demonstrated earlier, it takes a little over 1 second to completely write the ER3400.

In applications where the EAROM is "read-mostly" memory and is used to hold infrequently changing data or programs, slow erase and write times

are no problem. However, in instances where an EAROM is used to store data or programs during power-down conditions, slow write times are objectionable. The distinctive feature of the ER1711 is its ability to store its entire contents of the programmable memory in a single write pulse.

The static programmable-memory section of the ER1711 is addressed like any other system memory. It can hold constantly changing data with no restrictions on the number of read/write cycles. The contents of this memory can be transferred to the EAROM section through one write pulse. In this way, the EAROM can instantly save all data when the power fails. When power is returned, a simple pulse sequence restores the data from the EAROM to the programmable operating memory.

ER1711 Power Requirements

The ER1711 uses +5-V and −12-V power supplies for normal programmable-memory operation. These supplies must be kept constant

Figure 5: *A circuit that allows data stored in the programmable memory of the ER1711 to be saved into and recalled from the electrically alterable read-only-memory section. Charge-pumping circuits are used to generate the relatively high voltages needed for the erase, store, and recall cycles of the ER1711. Power-down and power-up states are initiated by the System Save and Data Recall signals.* →

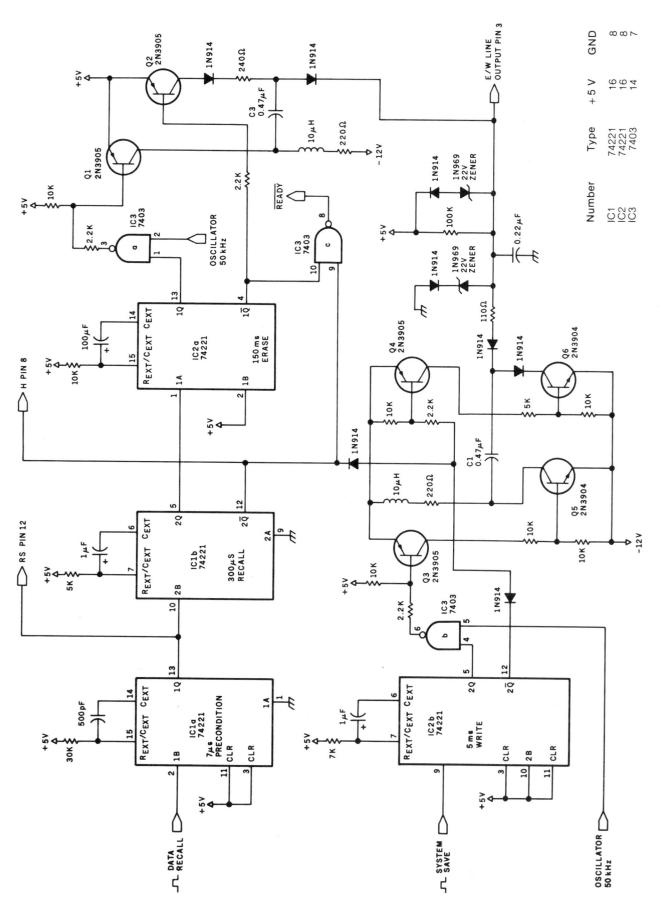

Figure 5: *A circuit that allows data stored in the programmable memory of the ER1711 to be saved into and recalled from the electrically alterable read-only-memory section. Charge-pumping circuits are used to generate the relatively high voltages needed for the erase, store, and recall cycles of the ER1711. Power-down and power-up states are initiated by the System Save and Data Recall signals.*

within a 5% tolerance. Other voltages used are as follows for EAROM operations:

$$V_W (-17 \text{ V to } -21 \text{ V})$$

is used to transfer data from programmable memory to EAROM;

$$V_R (-8 \text{ V to } -15 \text{ V})$$

is used to transfer data from EAROM to programmable memory; and

$$V_E (+25 \text{ V to } +30 \text{ V})$$

is used for erasing the EAROM storage cells.

Since these voltages are required only as single pulses during either power-up or power-down cycles, they do not require separate power supplies. All voltages can be generated by charge-pumping circuits operated from +5-V and −12-V supplies.

Figure 5 is the schematic diagram of a circuit which accomplishes the save-data and recall-data functions and maintains the proper timing and voltages required for the various operations.

Data Restoration

The static-programmable-memory portion of the ER1711 has an access time of 900 ns. Upon initial application of power to the system, if this programmable memory is to be loaded from EAROM, the computer signals this action by pulsing the *data-recall* input line. This triggers a 300µs ramped voltage to the erase/write (E/W) line of the ER1711. After 300 µs, the hold line (H) is brought high for 150 ms, and the E/W line is set at a potential of 27 V. At the conclusion of this sequence, the ER1711's programmable memory will contain the former contents of the EAROM, and the EAROM will be erased.

Saving the data stored in the programmable memory is started by raising the *system-save* line to a logic 1. The resultant application of a −20-V pulse of a 1-to-10-ms duration to the ER1711 causes the data to be stored in the electrically alterable read-only memory. Data retention time varies according to the pulse duration, from 3 days for a 1-ms pulse to 30 days for a 10-ms pulse.

In Conclusion

I do not expect that you will immediately convert your computer memory to EAROM, but at least you now know what EAROM is. In my own case, I have chosen to use ER1711 memory devices for my automotive computer. I can only speculate on the final configuration, but at least I can count on not having to be concerned with standby power consumption and battery backup.

For further information on EAROMs or determination of price and availability contact:

General Instrument Corp
600 W John St
Hicksville NY 11802

EAROM specifications and diagrams reprinted courtesy of General Instrument Corporation.

Operation of the ER1711

In normal operation the ER1711 operates as a programmable memory. Before powering down, the data can be stored in the electrically alterable, nonvolatile read-only-memory (EAROM) cells by a single, negative write pulse. When power is restored, the previously saved data can be recalled by a power-up and data-recall cycle which transfers all this saved data to the programmable memory. It is suggested that an erase cycle be performed soon after the data-recall cycle, so that the memory will be prepared in case of another power-down cycle. The EAROM cycles operate as explained here.

Erase Cycle:

1. *The H line should be high and the RS and CE lines low.*
2. *Positively pulse the E/W line to +25 V for 100 to 200 ms.*

Stored-Data Cycle:

1. *The H line should be held high and the RS and CE lines low.*
2. *Pulse the E/W line negative to -22 V for 1 to 10 ms to store nonvolatile data.*
3. *The nonvolatile EAROM memory cells must have been previously erased for valid data retention.*

Power-Up and Recall Cycle:

1. *Turn on power with CE and RS held to ground and the H and E/W lines based to V_{ss}.*
2. *When power is on, lower H and pulse RS to precondition the EAROM cells.*
3. *Lower RS and ramp down the E/W line at a rate of -0.1 V/µs. This can be done with a series resistor of 470 K ohms in the E/W line.*
4. *After the E/W recovery time, bring both the E/W and H lines to +V_{ss}.*
5. *If in the course of erasing data, the power shuts down again, the erase cycle can be terminated and a write cycle immediately started without loss of data.*

In normal programmable-memory operation, only the +5-V and -12-V power supplies are required. The erase-store and recall cycles require momentary high-voltage pulses to tunnel charges through the negative-metallic-oxide-semiconductor (NMOS) memory transistors. These higher voltages can be created from the +5-V and -12-V power supplies using the charge-pumping circuits shown in figure 5. This circuit will generate the sequence of RS, H and E/W pulses needed for power-down and up sequencing. The power-down and up cycles are initiated by System Save and Data Recall signals respectively. Figure 5 is a suggested circuit using standard transistor-transistor logic (TTL) parts; a similar circuit can be designed with complementary-metallic-oxide-semiconductor (CMOS) logic instead.

Computerize a Home

I anxiously glanced around the Circuit Cellar. Devoid of the usual sounds of the stereo or television, the equipment fans imparted a distinctly uneasy sensation of mechanical presence.

The room was totally dark except for a few pilot lights and a video display. There were no games, no fast-moving program listings; only a single line was written on the screen. In the dim luminescence I could barely distinguish the furniture from the bookcases. A little experience navigating in the dark would have been useful, but I opted for modern technology and reassuringly patted the flashlight in my pocket.

I pushed the button on my digital watch and noted the time. As it neared the prearranged hour, I turned instinctively to the terminal. Soon I'd know which of us was in control!

Almost immediately the display changed and printed out "AUTOMATIC CONTROL INITIATED." Simultaneously I could hear a high-pitched noise. It sounded almost like an insect chirp. There are no crickets down here; it must be a subharmonic. So far so good, but did it work?

"Steve, did you just blow a fuse?" My wife stood in the doorway and called down the stairs. It didn't bother *her* that there weren't any lights on. After all, if you blow a fuse, shouldn't the lights be off?

"The kitchen light went off and the bedroom light came on. Wait! The bedroom light just went off and the kitchen light came back on. Now they're both off."

I grinned in a way that only a Cheshire cat could appreciate. "Sorry, Joyce, just experimenting on the latest article." Chuckling softly, I continued. "I hope you don't mind, but the computer seems to have taken over."

"Can it make beds?" she replied.

I should have known that she wouldn't be taken in that easily. "OK, I'll tell the computer to keep its sphere of influence to the cellar. I'll let you know what the password is later."

As if by magic, the Circuit Cellar lights were activated. The test was successful.

Security Versus Control

Even though it may seem true at times, our house has not been taken over by a computer. I was simply testing the latest addition to my home control system.

In previous chapters of this book, I presented a series of articles on the construction of a home security system. (See "Build a Computer Controlled Security System for Your Home" Part 1, page 6; Part 2, page 16; Part 3, page 30.) This was not a theoretical dissertation. It was, in fact, an overview of the system installed in my house. The original concept was configured around a single-board 8085 system and designed primarily as an alarm controller. Even though it works (our house has never yet been broken into!), it has definite limitations.

Eventually I became dissatisfied with just having a super burglar alarm. It seemed a shame to dedicate all that hardware and expense to a function with such a limited capacity. The obvious step was to expand the concept to be a "home control" system where security is but one of many possible applications. To do this requires more memory; the single board has been replaced by a 26 K byte Z80-based computer with a video display. Operating in either high-level or assembly language, it is as adept at keeping the checking account straight as it is at scanning input ports searching for an intruder. Add to it the ability to activate and communicate with my large disk-based development system, and it is indeed a powerful tool.

The major difference between the two system concepts is the output control structure. As an alarm, the computer is strictly configured to scan and analyze a multitude of event inputs, such as door switches and motion sensors. Its decision process is immediate, but its output control is relatively limited. These generally consist of several lights, a siren, and an automatic phone dialer. Even in

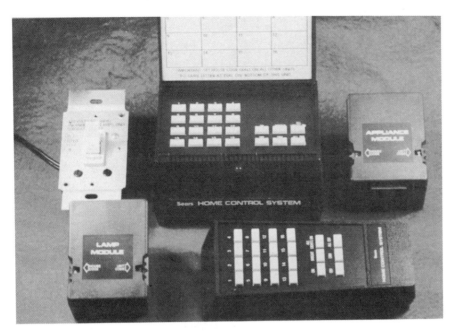

Photo 1: *BSR X-10 system as marketed by Sears and called the Sears Home Control System.*

Photo 2: *The internal electronics of the command-control console.*

the sophisticated system I presented, these hardwired outputs were kept to a minimum to reduce costs.

Generalized home control extends computer control capability far beyond the few outputs of the original system. It is conceivable that all of the lights and AC outlets in the house could be affected. A few lights outside are barely enough. Lighting in the bedrooms, kitchen, and garage should be included, with the stereo and television thrown in for good

measure. If you live in a cold climate and use an automobile engine-block heater, why not turn it on automatically before you get up in the morning? Tired of searching around in the dark for the light switch? Let the door sensor from the alarm system trigger the lights as you walk into a room. How about some soft music ten minutes after you enter? The list is endless.

This expansion seems to be a contradiction considering my previous

concern over wiring costs. To accomplish this feat, either every AC outlet must be directly wired to the computer through relays as in the original system *or* the control capability must be added remotely to each light and appliance.

AC Remote Control

This latter suggestion is not as farfetched as it might seem. There have been many technological advances in the past year. One of the more significant achievements comes from BSR (USA) Ltd—specifically in the area of AC remote control. The BSR X-10 control system is shown in photo 1. Clockwise from the center, the five components are: command console, appliance module, cordless controller, lamp module, and wall-switch module. With these units, low-cost AC control is a reality.

The BSR X-10, also marketed by Sears as the Sears Home Control System, operates through carrier current transmission from the command console to the receivers. When a button is pushed on the command console to activate a remote receiver, a coded signal is sent through the house wiring. Each receiver monitors these transmissions and responds only when its particular code is sent.

Figure 1a is a block diagram of the $39 command module and photo 2 shows its internal electronics. The heart of this, as well as the other system components, consists of custom large-scale integration (LSI) chips manufactured for BSR by General Instrument Corp. In normal operation the twenty-two-button keypad is continuously scanned. When a key is pressed, this designated function and a house code (previously set by a thumbwheel switch on the bottom of the command console) are combined into a single message. The digital message is directed to the transmitter section, where it modulates a 120 kHz carrier. The control signal appears on an oscilloscope as a series of pulse bursts. This is shown in photo 3.

There is a second method where the command console designates a control function and transmits a message. Each control console contains a ultrasonic receiver. In the picture this is the metallic cylindrical component with the two protruding pins and shielded cable soldered on them. The BSR X-10 system facilitates

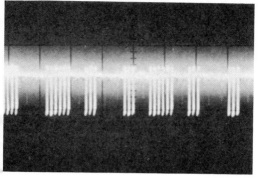

Photo 3: *Oscilloscope picture of command-control console transmission on the AC line. (Photo courtesy of Mark Scheffler.)*

Photo 4: *Handheld cordless controller showing top and internal circuitry.*

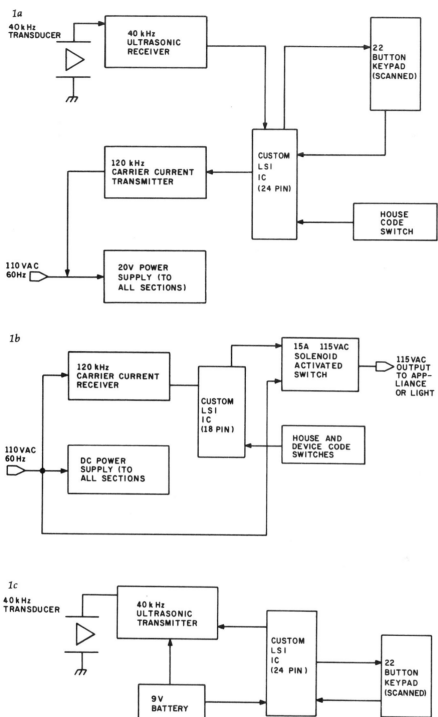

Figure 1: *Block diagrams of the integral parts of the BSR X-10. The block diagram for the control console is given in figure 1a, the appliance module in figure 1b, and the remote control transmitter in figure 1c.*

remote channel and function selection through a handheld ultrasonic transmitter. This unit is shown in photo 4 and diagrammed in figure 1c.

When a key is pressed, it is encoded and transmitted as a series of 40 kHz tone bursts. The command console, receiving this information through its ultrasonic receiver section, takes this data as if a button had been pushed on the command console. It then adds the house code and simultaneously transmits the command message over the house wiring.

The receiver part of the system is also quite sophisticated, considering that each receiver costs less than $15. These receivers, shown in photos 5 and 6, can be placed virtually anywhere. An overhead light can be accommodated by replacing the standard on/off wall switch with a wall-switch module. An appliance such as a dehumidifier is controlled through an appliance module.

All receivers are basically the same. A block diagram of an appliance module is shown in figure 1b. The receiver section monitors the AC line waiting for a coded message corresponding to its unique house (A thru P) and unit device (1 of 16) code.

To turn on channel 10, simply press "10" and then the "ON" button sequentially. When the appliance module activates, it sounds like a relay engaging. In actuality, BSR uses an inexpensive solenoid to operate a 15 A push-button Microswitch.

The lamp and wall-switch modules use a triac instead of this pseudo-relay. Unlike the appliance module, which only operates as an on/off switch, these units have the additional ability to automatically brighten or dim when the corresponding function buttons are pressed on the command console. Finally, all receivers can be locally activated without the command console. To turn on a light or motor, simply flip the power switch from on to off and

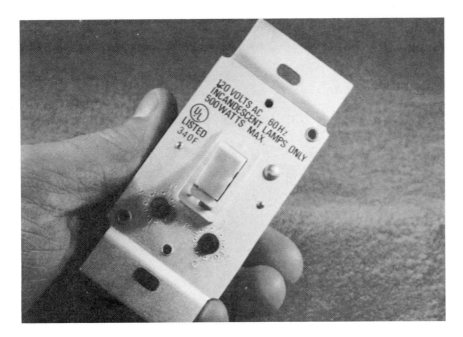

Photo 5: *The wall switch module replaces standard wall switch and allows remote control. Two slotted-top rotary switches under the switch lever are for setting house and device codes.*

Photo 6: *The appliance module.*

back to on again. This automatically triggers the receiver module into an on condition.

Controlling the BSR X-10

When I first started using the BSR X-10, I could hardly believe its versatility and low cost. The only problem is that operation of the BSR X-10 is completely manual. The only way to use the control receivers is through the command console or ultrasonic transmitter and by physically pressing the buttons.

I would not say that I have a never-say-die attitude, but considering my original security system, with an average cost of $250 per AC output channel, my future computer-controlled house depended heavily on less expensive input/output (I/O). It was absolutely necessary to find some method of utilizing the control receivers.

Three possible solutions came to mind:

● Directly synthesize the command-console waveform and transmit it directly onto the AC line.
● Brute force contact closure—attach either relays or complementary-metal-oxide semiconductor (CMOS) switches in parallel with the push buttons and activate the relays from the computer.
● Synthesize the waveform from the ultrasonic controller and let the computer "talk" to the command console.

Simulating the command-console output sounds simple in theory. (This is somewhat like estimating software costs.) Simulating the device-control code and using it to modulate a 120 kHz carrier frequency leads to contact with a hostile environment. The output from the computer must be attached to the AC line. This requires isolation through either transformers or optoisolators, plus many discrete components to properly match impedances. It is a shame to reinvent the wheel when BSR has already designed such an effective transmission system. Although possible in theory, this approach is too messy to warrant further consideration.

The second alternative is brute force. This can usually work, but you must be careful. In essence this method entails wiring relays or CMOS switches across the push buttons and remotely, but still mechanically or electronically, closing the contacts corresponding to a particular button. Figure 2a illustrates the keypad connections for both the command console and cordless controller. The configuration is a 3 by 8 scanning matrix. To turn on channel 6, simply short pins 28 and 18 together. Likewise, "dim" would be pins 25 and 23. While twenty-two separate single-pole, single-throw switches could be used, figure 2b demonstrates an easier alternative.

Two CMOS switches can be used in combination with the ultrasonic controller to provide this capability. Connected to 5 bits of a latched parallel output port, the two integrated circuits channel the appropriate lines together. To turn on channel 12, a row-select code of binary 001 would be set on B2, B1,

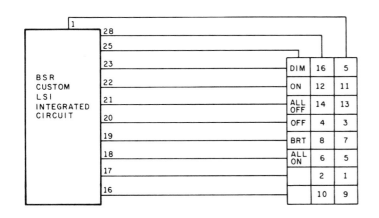

2a

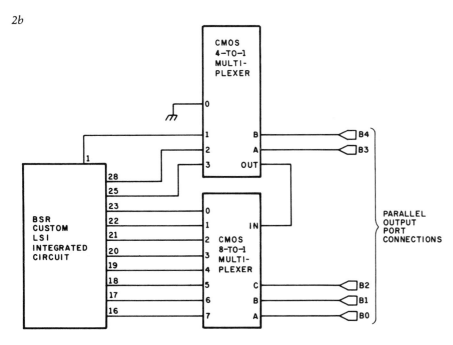

2b

Figure 2: *The keyboard in the BSR X-10 command console and remote control unit is connected to a custom LSI integrated circuit. Figure 2a shows the keyboard in relation to lines coming from the BSR custom integrated circuit. The functional schematic in figure 2b could be used to replace the keyboard with an I/O port.*

and B0, respectively, and a column-select code of binary 10 would be set on B4 and B3. The ON key would be a code of 11100 for bits B4 thru B0, for example.

The circuit of figure 2b will work only with the handheld battery-controlled unit. The command-console electronics, which run on −20 V, can use the same logical concept, but relays must be substituted for the CMOS switches. The command console is not isolated and its electronics are floating at 120 VAC. To be totally safe, it is best not to bother with it.

Hardwiring to the handheld unit keyboard will work, but it also has some detrimental features. In operation, the ultrasonic unit consumes an

average of 30 mA, while peak currents are about 100 mA. Alkaline batteries are a must. Short of direct connection to the computer's power supply through a 9 V regulator, there is always the hazard of battery brownout. If I were depending upon this system, I would not have a critical component powered by battery.

Talking to the BSR X-10

The sensible alternative is to construct an interface that facilitates cordless communication between the computer and the BSR X-10 command controller. Safety is the primary consideration. There is no hazard in using the controller or receivers as long as their cases are

intact. The BSR X-10 is Underwriters' Laboratories listed. Attachments between the computer and the command module must be done carefully and only by experienced people. By maintaining the structural integrity of the components, you are not limited to use with the computer. The command console can be moved around the house, and it is placed within range of the computer only when automatic control is desired.

Practical accomplishment of this goal is achieved using the ultrasonic receiver found within the command module. An interface is constructed that formats function codes into message strings; these strings are transmitted to the command console as 40 kHz pulses. In essence, the interface simulates the activity of a cordless-controller unit.

Figure 3 describes in detail the communication between the two subsystem components. Each of the twenty-two buttons has a unique 5-bit code (listed in table 1). For example, channel 5 has a code of 00010 with respect to bit positions D8, D4, D2, D1, and F. The ALL LIGHTS ON key generates the code 00011.

The actual message that communicates this selection is approximately 100 ms long and is composed of thirteen 8 ms segments. Each segment consists of a burst of 40 kHz directed to an ultrasonic transducer. Data is pulse-width modulated. A logic 1 is a 4 ms burst and a logic 0 is a 1.2 ms burst.

To signify channel 5, the interface first sends a start bit to alert the receiver of the pending message transmission. This is a 40 kHz tone for 4 ms. Next, the 5-bit selection code is sequentially transmitted as a series of 1.2 and 4 ms bursts of 40 kHz. This is followed by transmission of the logical inversion of the 5-bit selection code and a 16 ms end-of-message tone. All messages use the same format; only the 5-bit selection code varies.

Figure 4 is an interface specifically designed to send this message and facilitate wireless remote control. Incorporating complete circuitry for address decoding and data storage, it appears to the computer as a single output port. Turning on the table lamp is as simple as sending a 1-byte output to the interface port. As with

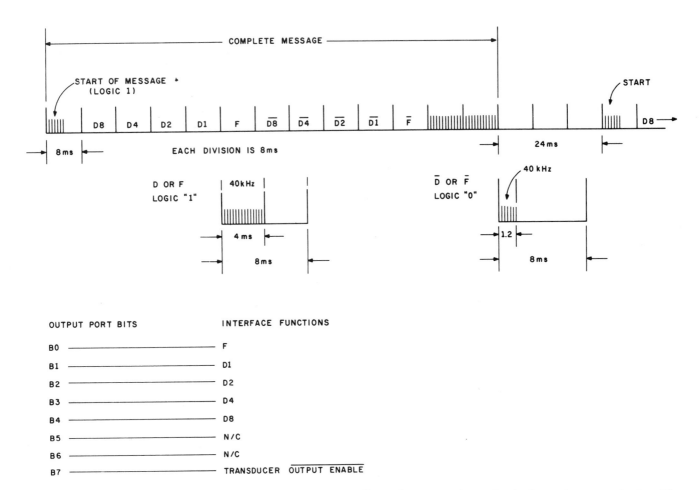

Figure 3: *Description of coded message sent from the cordless controller to the command console via ultrasonic communication. The necessary codes are shown in table 1.*

the majority of my designs, it is both processor and speed independent. It works equally well in BASIC or assembly language programs. Connected to port 9 (as in my example software), turning on a table lamp or the hall lights in BASIC is a one-line command, OUT 9,5 (from the code list of table 1). Turning it off is simply OUT 9,7.

The circuit will work on virtually any computer, although the pin designations in figure 4 refer specifically to the Radio Shack TRS-80 Model I. All connections are made directly to the computer address and data buses. In the TRS-80 this is done through the expansion connector. In a computer such as an Apple II, the circuit could be built to plug directly into the back-plane connector or to be connected by a ribbon cable.

The electronics can be divided into three subsystems: port latch and address decoding, pseudo pulse-width modulator, and message serializer. Photo 7 illustrates the prototype of figure 4.

ICs 9 thru 13 make up the address decoding and data latch. For a TRS-80, I have arbitrarily chosen an address of 127 decimal (in my software examples, I use port 9). When

the address bus and OUT line (corresponding to I/O WRITE on some systems) indicate execution of an output command, the contents of the data bus are stored in ICs 11 and 12.

CHANNEL NUMBER OR FUNCTION	BINARY CODE D8 D4 D2 D1 F					DECIMAL OUTPUT TO INTERFACE CIRCUIT
1	0	1	1	0	0	12
2	1	1	1	0	0	28
3	0	0	1	0	0	4
4	1	0	1	0	0	20
5	0	0	0	1	0	2
6	1	0	0	1	0	18
7	0	1	0	1	0	10
8	1	1	0	1	0	26
9	0	1	1	1	0	14
10	1	1	1	1	0	30
11	0	0	1	1	0	6
12	1	0	1	1	0	22
13	0	0	0	0	0	0
14	1	0	0	0	0	16
15	0	1	0	0	0	8
16	1	1	0	0	0	24
ALL OFF	0	0	0	0	1	1
ALL LIGHTS ON	0	0	0	1	1	3
ON	0	0	1	0	1	5
OFF	0	0	1	1	1	7
DIM	0	1	0	0	1	9
BRIGHT	0	1	0	1	1	11

Table 1: *Cordless controller push-button codes and decimal equivalents.*

Bits 0 thru 4 will contain the function code (from table 1) and bit 7 is used to turn the transmitter output on and off. For further information on address decoding and output ports, I refer you to *Ciarcia's Circuit Cellar, Vol. I* from BYTE Books and the article entitled "Memory Mapped I/O," which first appeared in the November 1977 *BYTE*, page 10.

In figure 4, the 5-bit function code, as well as its logical inversion, are attached to a 16-to-1 multiplexer, IC1. As the 4-bit counter IC7 increments, each of the input lines of the multiplexer is sequentially routed to the output, pin 10. With address position 0 permanently tied high and the next ten addresses wired as function-code inputs, the output of IC1 will reflect the first eleven 8 ms message segments.

ICs 3, 5, 6, and 8 act as a digital modulator. If the output of IC1 pin 10 is a logic 1 (such as the start bit), a 4

Photo 7: *Prototype of the circuit shown in figure 4. The ultrasonic transducer is remotely located, and ICs 9, 10, 11, 12, and 13 are contained on another board.*

As of the writing of this article, Mountain Hardware Inc (300 Harvey West Blvd, Santa Cruz CA 95060) has announced a plug-in card for the Apple II that, like the control card described in this article, transmits to the BSR X-10 Command Console. In addition, the company offers control software tailored to the Apple II with at least 32 K bytes of programmable memory. Cost of the unit is $200 for the controller board alone and $300 for the controller board, the X-10 Command Console, and three remote modules.

The following items are available from:
   The Micromint Inc
   917 Midway
   Woodmere NY 11598
   Telephone: (516) 374-6793

1. *Assembled and tested interface board with manual, power supply, case, and cable:*
   For TRS-80 Model I $109.95
   For Apple II $114.95
   For S100 $119.95
2. *Real-time control software for TRS-80* $19.95
3. *40 KHZ transducer* $6.50
   All prices are valid until Dec. 31, 1981; call for prices after that date. New York residents please add 7% sales tax.

ms burst of 40 kHz will be routed through IC5 and appear at pin 6. A logic 0 on pin 10 results in a 1.2 ms burst. The timing of these events is rather critical. The rate of clock 1 (IC8d) should be as close to 125 Hz as possible (8 ms period), and clock 2 (IC8e and IC8f) should be similarly set to 40 kHz. Use potentiometer R1 to set the monostable multivibrator (or one-shot) IC6a to a period of 4 ms. Use R2 to set the one-shot IC6b to 1.2 ms.

The output of IC5 should generate the first eleven segments of the message. IC2, using the same technique as IC1, adds a 16 ms end-of-message tone burst as segments 12 and 13. The message is repeated in 24 ms as the counter (IC7) loops to 0. It will send the same data as long as the contents of ICs 11 and 12 have not changed and the output-enable line has not been brought high.

All of the components (except possibly the 40 kHz transducers) are readily available. Low-power Schottky transistor-transistor logic (TTL) devices should be used where specified to properly interface with the TRS-80 or similar low-power bus systems.

One further note for prospective TRS-80 circuit builders. To use this interface properly, you must have Level 2 BASIC to address output ports. Also, in most Level 2 systems,

+5 V on the expansion connector has been disconnected at the factory. It will be necessary, therefore, to provide a separate 5 V 300 mA power supply for the interface electronics.

Using the Interface

A typical application is demonstrated in figure 5. The receivers can be placed around the home to control a variety of appliances and lights. With the addition of the real-time clock outlined in a previous Circuit Cellar article ("Anyone Know the Real Time?", page 80) you can add timed activation of these control functions as well.

Listing 1 shows a simple BASIC program that demonstrates the interface capabilities. The command console is plugged in and positioned within 20 feet on a direct unobstructed line with the interface output transducer. The program starts by asking if you want to clear all outputs and start fresh. Since the BSR X-10 is an open-loop control system, and you have no way of knowing which receivers are activated, this is a prudent choice.

To turn on channel 6, simply answer the appropriate questions with "6" and "ON." The status of all channels can be reviewed at any time.

The program responds by calling a control output routine. Turning channel 6 on requires two outputs to

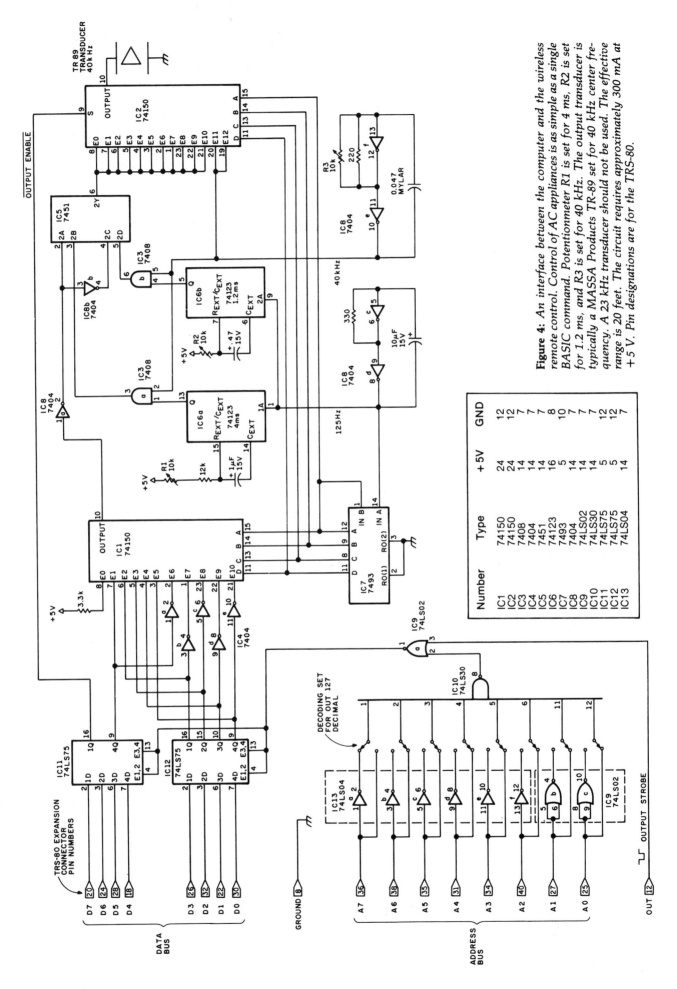

Figure 4: *An interface between the computer and the wireless remote control. Control of AC appliances is as simple as a single BASIC command. Potentiometer R1 is set for 4 ms, R2 is set for 1.2 ms, and R3 is set for 40 kHz. The output transducer is typically a MASSA Products TR-89 set for 40 kHz center frequency. A 23 kHz transducer should not be used. The effective range is 20 feet. The circuit requires approximately 300 mA at +5 V. Pin designations are for the TRS-80.*

Number	Type	+5V	GND
IC1	74150	24	12
IC2	74150	24	12
IC3	7408	14	7
IC4	7404	14	7
IC5	7451	14	7
IC6	74123	16	8
IC7	7493	5	10
IC8	7404	14	7
IC9	74LS02	14	7
IC10	74LS30	14	7
IC11	74LS75	5	12
IC12	74LS75	5	12
IC13	74LS04	14	7

the command console. One sets channel 6 (as if pressing the 6 button), and the other sets the "on" function (as if pressing the ON button). To allow enough time for the command console to respond, delay loops are inserted. The result is a 2-second signal to set device code 6 and a 2-second message that tells it to turn on. The process can be reversed with a 6 and an OFF program command. All sixteen channels can be just as easily cycled.

Listing 2 is the logical extension of this basic concept. Using a real-time clock, you can create a list of precisely timed events. It can be used to control house lighting during vacations or to turn the coffee maker on at 6:30 AM. The program incorporates a default list of data statements. Each statement is formatted as time, channel, and function. To turn on a coffee maker connected to a unit 6 appliance module simply write: DATA 0630,6, "ON." This technique allows us to set

up a specific vacation or holiday repertoire. Just load the program and run it. The list of control data can be added to while the program is running. This allows specific actions such as shutting off the television at 2 AM in case you fall asleep during the late show. Entries such as these are retained only as long as the program is running. They must be reentered if the BASIC program is terminated.

Conclusion

I always try to present interfaces and applications that I think will interest *BYTE* readers. I consider this one to be particularly significant considering the cost advantages over earlier technology. I will not replace the relay-controlled lighting in my home, but further expansion of AC control will use the hardware from this article. There are, of course, many situations where the BSR X-10 is inappropriate, but considering the sophistication when it is connected to a computer, I am going to look a lot harder for ones that apply.

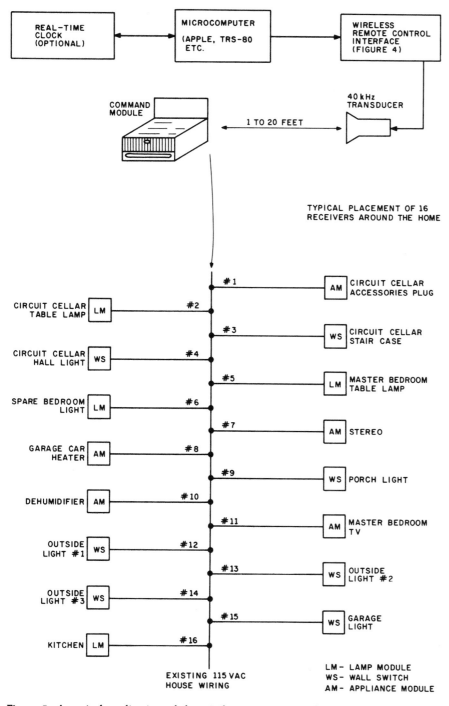

Figure 5: *A typical application of the wireless remote control. All of the modules are connected over the house wiring.*

Listing 1: *Demonstration program for the Sears Home Control System.*

```
LIST

90 REM THIS PROGRAM ALLOWS DIRECT COMPUTER CONTROL OVER THE SEARS HOME CONTROL SYSTEM
92 REM
94 REM COPYRIGHT 1979  STEVEN A. CIARCIA
96 REM
98 REM
100 REM SET UP TABLE OF CHANNEL/OUTPUT CODES
105 DIM C(20),S(20)
110 DATA 12,28,4,20,2,18,10,26,14,30
115 DATA 6,22,0,16,8,24
120 FOR X=1 TO 16
125 READ C(X) :REM C(X) IS CHANNEL NUMBER
130 NEXT X
135 GOSUB 200 :GOTO 300
190 REM
195 REM
200 PRINT "CURRENT STATUS IS :"
205 FOR X=1 TO 16
210 PRINT "CHANNEL ";X;" IS   "; :IF S(X)=1 THEN PRINT"ON" ELSE PRINT"OFF"
220 NEXT X
225 RETURN
300 PRINT "DO YOU WANT TO CLEAR ALL OUTPUTS TO START":INPUT A$
305 IF A$="YES" THEN F=1 :GOSUB 9050:FOR Z=1 TO 16:S(Z)=0 :NEXT Z :REM CLEAR BSR OUTPUTS
310 PRINT "SET CONTROLLER OUTPUTS BY ENTERING CHANNEL NO. AND FUNCTION"
315 PRINT"ENTER CHANNEL NO.      (0 TO EXIT) ";:INPUT C
317 IF C=0 THEN GOSUB 200 :GOTO 300
320 PRINT"CHANNEL ";C;" IS   ";:IF S(C)=1 THEN PRINT"ON" ELSE PRINT"OFF"
325 PRINT"ON,OFF,NEXT,OR REVIEW ;":INPUT A$
330 IF A$="ON" THEN S(C)=1:X=C:GOSUB 9000:F=5:GOSUB 9050 :GOTO 315:REM TURN CHANNEL C ON
340 IF A$="OFF" THEN S(C)=0:X=C:GOSUB 9000: F=7:GOSUB 9050 :GOTO 315:REM TURN CHANNEL C OFF
350 IF A$="NEXT" THEN C=C+1:GOTO 320
360 GOSUB 200 :GOTO 315
8996 REM
8998 REM
9000 REM BSR HOME CONTROL DRIVER
9010 REM C(X) IS CHANNEL CODE
9015 OUT 9,C(X) :REM SET CHANNEL
9020 GOSUB 9085
9025 RETURN
9050 REM FUNCTION DRIVER
9060 REM F=FUNCTION CODE
9065 OUT 9,F
9070 GOSUB 9085
9075 OUT 9,128 :REM BIT 7 SHUTS OFF TRANSDUCER OUTPUT
9080 RETURN
9082 REM
9084 REM
9085 FOR Q=0 TO 900 :NEXT Q :RETURN :REM DELAY TIMER

READY




RUN


CHOOSE ONE OF THE FOLLOWING :

1.   AUTOMATIC CONTROL SYSTEM ON
2.   MANUAL CONTROL / CURRENT STATUS
3.   PRINT THE CURRENT TIME
4.   REVIEW DEFAULT SETTINGS AND ADD TO CONTROL LIST

 YOUR CHOICE ? 4

DO YOU WANT TO REVIEW THE DEFAULT SETTINGS  (Y/N) ? Y
  1 .AT   2  HOURS   0  MINUTES     TURN CHANNEL  10   ON
  2 .AT  17  HOURS   0  MINUTES     TURN CHANNEL  10   OFF
  3 .AT  18  HOURS  30  MINUTES     TURN CHANNEL   6   ON
  4 .AT  19  HOURS  25  MINUTES     TURN CHANNEL   6   OFF
  5 .AT  19  HOURS  40  MINUTES     TURN CHANNEL   6   ON
  6 .AT  20  HOURS  20  MINUTES     TURN CHANNEL   6   OFF
  7 .AT  20  HOURS  35  MINUTES     TURN CHANNEL   6   ON
  8 .AT  21  HOURS  50  MINUTES     TURN CHANNEL   6   OFF
  9 .AT  22  HOURS   0  MINUTES     TURN CHANNEL   6   ON
 10 .AT  23  HOURS  50  MINUTES     TURN CHANNEL   6   OFF
 11 .AT   1  HOURS  50  MINUTES     TURN CHANNEL   6   ON
 12 .AT   2  HOURS  45  MINUTES     TURN CHANNEL   6   OFF
```

```
13 .AT  16  HOURS   0  MINUTES    TURN CHANNEL  1  ON
14 .AT  22  HOURS   0  MINUTES    TURN CHANNEL  1  OFF
15 .AT  23  HOURS   0  MINUTES    TURN CHANNEL  4  ON
16 .AT   0  HOURS  30  MINUTES    TURN CHANNEL  4  OFF
17 .AT  19  HOURS  30  MINUTES    TURN CHANNEL  5  ON
18 .AT  21  HOURS  20  MINUTES    TURN CHANNEL  5  OFF
19 .AT  22  HOURS   0  MINUTES    TURN CHANNEL  5  ON
20 .AT   1  HOURS   0  MINUTES    TURN CHANNEL  5  OFF

        1......CHANGE LIST
        2......ADD TO LIST
        0......EXIT TO MENU
? 2
  ENTER TIME ,CHANNEL, AND FUNCTION
ENTRY NO. 21  ? 2330,3,ON
ENTRY NO. 22  ? 0,0,0

        1......CHANGE LIST
        2......ADD TO LIST
        0......EXIT TO MENU
? 0


CHOOSE ONE OF THE FOLLOWING :

1.  AUTOMATIC CONTROL SYSTEM ON
2.  MANUAL CONTROL / CURRENT STATUS
3.  PRINT THE CURRENT TIME
4.  REVIEW DEFAULT SETTINGS AND ADD TO CONTROL LIST

  YOUR CHOICE ? 1
AUTOMATIC CONTROL INITIATED
 23 HOURS  43 MINUTES
```

Listing 2: *Program to compare the time from a real-time clock against a list of operations to be performed at specific times. A sample run of the program demonstrates how the entries may be varied.*

```
RUN
CURRENT STATUS IS :
CHANNEL   1  IS  OFF
CHANNEL   2  IS  OFF
CHANNEL   3  IS  OFF
CHANNEL   4  IS  OFF
CHANNEL   5  IS  OFF
CHANNEL   6  IS  OFF
CHANNEL   7  IS  OFF
CHANNEL   8  IS  OFF
CHANNEL   9  IS  OFF
CHANNEL  10  IS  OFF
CHANNEL  11  IS  OFF
CHANNEL  12  IS  OFF
CHANNEL  13  IS  OFF
CHANNEL  14  IS  OFF
CHANNEL  15  IS  OFF
CHANNEL  16  IS  OFF
DO YOU WANT TO CLEAR ALL OUTPUTS TO START
? YES
SET CONTROLLER OUTPUTS BY ENTERING CHANNEL NO. AND FUNCTION
ENTER CHANNEL NO.     (0 TO EXIT) ? 2
CHANNEL  2  IS  OFF
ON,OFF,NEXT,OR REVIEW ;
? ON
ENTER CHANNEL NO.     (0 TO EXIT) ? 5
CHANNEL  5  IS  OFF
ON,OFF,NEXT,OR REVIEW ;
? ON
ENTER CHANNEL NO.     (0 TO EXIT) ? 2
CHANNEL  2  IS  ON
ON,OFF,NEXT,OR REVIEW ;
? NEXT
CHANNEL  3  IS  OFF
ON,OFF,NEXT,OR REVIEW ;
? REVIEW
CURRENT STATUS IS :
CHANNEL   1  IS  OFF
```

```
CHANNEL  2  IS  ON
CHANNEL  3  IS  OFF
CHANNEL  4  IS  OFF
CHANNEL  5  IS  ON
CHANNEL  6  IS  OFF
CHANNEL  7  IS  OFF
CHANNEL  8  IS  OFF
CHANNEL  9  IS  OFF
CHANNEL  10  IS  OFF
CHANNEL  11  IS  OFF
CHANNEL  12  IS  OFF
CHANNEL  13  IS  OFF
CHANNEL  14  IS  OFF
CHANNEL  15  IS  OFF
CHANNEL  16  IS  OFF
ENTER CHANNEL NO.     (0 TO EXIT) ?


LIST

100 REM
110 REM THIS PROGRAM PROVIDES REAL TIME CONTROL OF AC APPLIANCES
120 REM BY CONNECTING THE SEARS HOME CONTROL SYSTEM AND A REAL TIME CLOCK
130 REM TOGETHER.
140 REM
150 REM COPYRIGHT 1979   STEVEN CIARCIA
160 REM
170 REM
180 GOSUB 350 :REM LOAD DATA TABLES
190 REM PROGRAM OPTIONS ARE MADE THROUGH MENU SELECTIONS
200 PRINT :PRINT: PRINT "CHOOSE ONE OF THE FOLLOWING :"
210 PRINT
220 PRINT"1.   AUTOMATIC CONTROL SYSTEM ON"
230 PRINT"2.   MANUAL CONTROL / CURRENT STATUS
240 PRINT"3.   PRINT THE CURRENT TIME"
250 PRINT"4.   REVIEW DEFAULT SETTINGS AND ADD TO CONTROL LIST"
260 PRINT
270 PRINT" YOUR CHOICE "; :INPUT Z1
280 IF Z1=1 THEN PRINT"AUTOMATIC CONTROL INITIATED" : GOTO 1190
290 IF Z1=2 THEN GOSUB 740 :GOTO 810
300 IF Z1=3 THEN GOSUB 1470 :PRINT :PRINT"THE PRESENT TIME IS "; :GOSUB 1130 :GOTO 200
310 IF Z1=4 THEN GOTO 940
320 GOTO 200
330 REM
340 REM
350 REM SET UP TABLE OF CHANNEL/OUTPUT CODES
360 DIM C(20),S(50)
370 DATA 12,28,4,20,2,18,10,26,14,30
380 DATA 6,22,0,16,8,24
390 FOR X=1 TO 16
400 READ C(X) :REM C(X) IS CHANNEL NUMBER
410 NEXT X
420 REM WHEN PROGRAM IS INITIATED THE FOLLOWING DATA TABLE CONSTITUTES THE DEFAULT CONTROL SETPOINTS
430 REM SETPOINTS ARE STORED AS DATA STATEMENTS IN THE FORM OF TIME,CHANNEL,AND FUNCTION
440 REM W=TOTAL NUMBER OF DATA STATEMENTS
450 REM
460 DIM W(50),A(50),B(50),A$(50),L(50)
470 W=20 :REM W=TOTAL NUMBER OF DEFAULTS
480 DATA 0200,10,"ON" :REM DEHUMIDIFIER
490 DATA 1700,10,"OFF"
500 DATA 1830,6,"ON" :REM SPARE BEDROOM LIGHTS
510 DATA 1925,6,"OFF"
520 DATA 1940,6,"ON"
530 DATA 2020,6,"OFF"
540 DATA 2035,6,"ON"
550 DATA 2150,6,"OFF"
560 DATA 2200,6,"ON"
570 DATA 2350,6,"OFF"
580 DATA 0150,6,"ON"
590 DATA 0245,6,"OFF"
600 DATA 1600,1,"ON" :REM CIRCUIT CELLAR ACCESS. PLUG
610 DATA 2200,1,"OFF"
620 DATA 2300,4,"ON" :REM CELLAR HALL
630 DATA 0030,4,"OFF"
640 DATA 1930,5,"ON" :REM MASTER BEDROOM
650 DATA 2120,5,"OFF"
660 DATA 2200,5,"ON"
670 DATA 0100,5,"OFF"
680 FOR L=1 TO W :READ A(L),B(L),A$(L) :REM SET TIME,CHANNEL,FUNCTION
```

```
690 NEXT L
700 RETURN
710 STOP
720 REM
730 RE197 REM
740 PRINT "CURRENT STATUS IS :"
750 FOR X=1 TO 16
760 PRINT "CHANNEL ";X;" IS   "; :IF S(X)=1 THEN PRINT"ON" ELSE PRINT"OFF"
770 NEXT X
780 RETURN
790 REM
800 REM
810 PRINT "DO YOU WANT TO CLEAR ALL OUTPUTS TO START":INPUT A$
820 IF A$="YES" THEN F=1 :GOSUB 1380:FOR Z=1 TO 16:S(Z)=0 :NEXT Z :REM CLEAR BSR OUTPUTS
830 PRINT "SET CONTROLLER OUTPUTS BY ENTERING CHANNEL NO. AND FUNCTION"
840 PRINT"ENTER CHANNEL NO.     (0 TO EXIT) ";:INPUT C
850 IF C=0 THEN GOTO 200
860 PRINT"CHANNEL ";C;" IS   ";:IF S(C)=1 THEN PRINT"ON" ELSE PRINT"OFF"
870 PRINT"ON,OFF,NEXT,OR REVIEW ;":INPUT A$
880 IF A$="ON" THEN S(C)=1:X=C:GOSUB 1330:F=5:GOSUB 1380 :GOTO 840:REM TURN CHANNEL C ON
890 IF A$="OFF" THEN S(C)=0:X=C:GOSUB 1330: F=7:GOSUB 1380 :GOTO 840:REM TURN CHANNEL C OFF
900 IF A$="NEXT" THEN C=C+1:GOTO 860
910 GOSUB 740 :GOTO 840
920 REM
930 REM
940 PRINT: PRINT"DO YOU WANT TO REVIEW THE DEFAULT SETTINGS  (Y/N) ";:INPUT B$
950 IF B$<>"Y" THEN GOTO 1000
960 FOR L=1 TO W
970 L1=INT(A(L)/100)
980 PRINT L;",";"AT ";L1;" HOURS ";A(L)-L1*100;" MINUTES    ";"TURN CHANNEL ";B(L);" ";A$(L)
990 NEXT L
1000 PRINT:PRINT"     1......CHANGE LIST":PRINT"     2......ADD TO LIST"
1010 PRINT"     0......EXIT TO MENU" :INPUT Z2
1020 IF Z2=0 THEN GOTO 200
1030 IF Z2=1 THEN PRINT"RECORD ENTRY TO BE CHANGED"; ELSE 1080
1040 INPUT Z3
1050 PRINT" PRESENTLY  ";A(Z3),B(Z3),A$(Z3)
1060 INPUT "TIME ,CHANNEL,AND ON OR OFF ";A(Z3),B(Z3),A$(Z3) :IF A(Z3)=0 THEN 1000
1070 PRINT "CHANGE ANOTHER  Y/N "; :INPUT Z$ :IF Z$="Y" THEN GOTO 1030 ELSE 1000
1080 IF Z2<>2 THEN GOTO 1000
1090 REM START ADDITIONS AT END OF DEFAULT LIST
1100 PRINT" ENTER TIME ,CHANNEL, AND FUNCTION"
1110 W=W+1 :PRINT"ENTRY NO."; W,:INPUT A(W),B(W),A$(W) :IF A(W)=0 THEN W=W-1 :GOTO 1000
1120 GOTO 1110
1130 REM 4 DIGIT FORMAT ROUTINE
1140 T2=H1*10+H0 :T3=M1*10+M0
1150 PRINT T2;"HOURS ";T3;"MINUTES" :RETURN
1160 REM
1170 REM
1180 REM
1190 REM CONTROL OUTPUT SUBROUTINE-----SETPOINT MONITOR
1200 T5=0
1210 GOSUB 1470  :REM GET TIME
1220 IF T1<>T5 THEN GOSUB 1130 :T5=T1 :REM PRINT TIME
1230 FOR L=1 TO W
1240 IF T1=A(L) THEN X=B(L) :GOSUB 1330 :GOSUB 1280
1250 NEXT L
1260 IF INP(0)<>0 THEN GOTO 200 :REM CHECK KEYBOARD FOR INTERRUPT INPUT
1270 GOTO 1210
1280 IF A$(L)="ON" THEN F=5:S(B(L))=1   :GOSUB 1400·
1290 IF A$(L)="OFF" THEN F=7 :S(B(L))=0 :GOSUB 1400
1300 RETURN
1310 REM
1320 REM
1330 REM BSR HOME CONTROL DRIVER
1340 REM C(X) IS CHANNEL CODE
1350 OUT 9,C(X) :REM SET CHANNEL
1360 GOSUB 1460
1370 RETURN
1380 REM FUNCTION DRIVER
1390 REM F=FUNCTION CODE
1400 OUT 9,F
1410 GOSUB 1460
1420 OUT 9,128 :REM BIT 7 SHUTS OFF TRANSDUCER OUTPUT
1430 RETURN
1440 REM
1450 REM
1460 FOR Q=0 TO 900 :NEXT Q :RETURN :REM DELAY TIMER
```

```
1470 REM THIS ROUTINE IS THE REAL TIME CLOCK INTERFACE DRIVER
1480 REM HARDWARE DESCRIBED IN AUG. '79 BYTE
1490 REM IT READS IN 2400 HR. FORMAT AND IS CONNECTED TO PORT 8
1500 RO=0 :OUT 3,254 :REM TURN ON PANEL LIGHT
1510 OUT 8,1 :OUT 8,0 :T=INP(8) :D=T AND 16
1520 IF D=16 THEN 1530 ELSE 1510
1530 MO=T AND 15 : GOSUB 1630
1540 M1=T AND 15 :GOSUB 1630
1550 HO=T AND 15 :GOSUB 1630
1560 H1=T AND 15 :GOSUB 1630
1570 T1=(H1*10+HO)*100+(M1*10+MO) :REM TIME IN 2400 FORMAT
1580 IF RO=T1 THEN 1590 ELSE RO=T1 :GOTO 1510
1590 OUT 3,255 :REM TURN OFF PANEL LIGHT
1600 RETURN
1610 REM T1=TIME    H1=TENS OF HOURS    HO=HOURS
1620 REM M1=TENS OF MINUTES    MO=MINUTES
1630 OUT 8,1 :OUT 8,0 :T=INP(8) :RETURN

READY
```

ANOTHER WAY TO COMPUTERIZE A HOME

Steve Ciarcia's article "Computerize a Home," which deals with utilizing the BSR X-10 Home Control System more fully by adding computer control, blazes a trail of interest to many. And his tracking of the amazing drop in system cost provided by the BSR technology is very graphic.

Readers of BYTE should be aware that some of the BSR command units do not include the microphone circuitry needed to accept the acoustic signals from the remote controller or Steve's interface. The command unit Model X10-014311, probably sold primarily as part of the $89 starter system, does not have the microphone. If you plan to implement Steve's approach, you must use the Model X10-014301.

Steve listed and evaluated the principle interface methods available between the X-10 and the computer. I think this area might deserve further review, especially in the light of the figure and caption. The principle options are:

1) Directly synthesize the command console waveform and impress it directly onto the AC line.
2) Brute-force contact closure—attaching computer-controlled relays or switches in parallel with the existing switches of the command unit.

3) Synthesize the waveform from the ultrasonic controller and let the computer "talk" to the command console.
4) In addition, synthesize an electrical waveform and inject it into the command console, bypassing the acoustic elements.

Rather than dismiss option 1 and ignore option 4, one might want to evaluate the choices on more substantive grounds, which might include the capabilities of the experimenter. Radio Shack sold a novice-level, carrier-current intercom kit for years which dealt with the "hostile" 110 VAC environment Steve worries about.

I opted for option 1, for two reasons: simplicity and cost. The hardware actually requires fewer discrete parts than Steve's design and eliminates all but two integrated circuits, an opto-isolator, and a 555 timer. Even more interestingly, I used the computer, not special hardware, to generate the waveforms. For these off/on-type waveforms, the computer is in its glory. Both the actual cost of parts and the time required to implement the hardware were less than one-half of Steve's cost. Further, I don't have to tie up or share a $50 command console.

I didn't explore option 4, but the trade-off between the cost of the acoustic transducer and opening the command unit probably favors option 3 for a transducer cost under $10.

In developing my software, I followed the structured programming approach because of two things I had in mind. I didn't want to dedicate a $1200 Apple II computer to the menial task of controlling a dozen light circuits, and I didn't want to reload and reinitialize the home-control program after each time I wanted to use the machine for something else. Because of this, my program is strictly modular and can be run in two modes: the interrupt mode where the home-control program runs continuously in background leaving the foreground available continuously for other uses (a very elementary time-share system), or in the alternate mode where home-control execution can be halted temporarily to make the machine available for other uses. Following this use, the home-control program will play "catch-up" in case any event times occurred while it was off-line.

To accomplish the above, I partitioned the modules of the program into two portions: that portion required to be in the computer's memory for program operation (the event-controlling program) and that portion required to interface with the human operator and allow changes, etc (the driver program). The event-controlling program (including the machine-language waveform-generator routine) occupies less than 3 K bytes of memory and is located at the high end of memory (with HIMEM set below it). With HIMEM set below it, the computer can be used normally; the BASIC commands RUN, LOAD, SAVE, NEW, etc can be used without erasing or corrupting the event-controlling program. The driver program is loaded when necessary to make changes.

I do believe implementing this approach is one step further along the path toward an economical, utilitarian use for a home computer.

Jim Fulton

ULTRASONIC SUBSTITUTION

The schematic diagram of figure 4 in "Computerize a Home," by Steve Ciarcia specifies that a Model TR-89 40 kHz ultrasonic transducer from Massa Products Corporation be used. Several readers have asked how to get this component.

Steve Ciarcia suggests that an equivalent transducer from Panasonic be substituted for the Massa Products unit. The Panasonic transducer may be ordered from: The MicroMint Inc, 917 Midway, Woodmere NY 11598, telephone (516) 374-6793.

The MicroMint stock number for the device is MM1002; the cost is $6 postpaid.

THE VERY BUSY BOX

Dear Steve,

Thank you for writing that great article about the BSR X-10 controller. I am interested in designing an interface for the Heath H8 computer, but would like to avoid using analog circuitry, wherever possible, by deriving the timing from a single clock signal. Obviously, not many of the frequencies called for in your article are going to be met exactly. Could you give me some idea of the tolerances involved?

John R Souvestre

In that article, I was careful to quote only the manufacturer's specifications, the time intervals 1.2 ms and 4.0 ms, and the frequency 40 kHz. Experimentation using an oscilloscope and a frequency counter has demonstrated a fairly large tolerance in these settings. I have been successful with settings in the ranges of 1.1 ms to 1.7 ms, 3.5 ms to 4.5 ms, and 39,500 Hz to 40,800 Hz, respectively. The amount is dependent upon the particular command controller.

If you really want bare bones, use an NE555 timer as a 40 kHz oscillator and gate its output with

*software timing....***Steve***

Dear Steve,

I have been considering methods to control most of the appliances in my house through my computer, and your article has given me the means. I have started collecting parts, but I ran into a problem: I cannot find the address of Massa Products, supplier of the ultrasonic transducer.

Mike McLennan

*Unfortunately, Massa Products has a minimum-order policy, so it may not be cost-effective to use their transducer unless you need other parts too. You can use a Panasonic EFROSB40K2, available from Panasonic's distributors, or as part number MM1002 from: The MicroMint, 917 Midway, Woodmere NY, 11598, (516) 374-6793; cost: $6....***Steve***

Dear Steve,

Your article about computer control of appliances prompted me to purchase a BSR X-10 system. I have devised a simple remote audio-volume control which uses a lamp-dimmer module, acting on a photoresistive cell. By inserting the cell in line between the preamp and power amp of my stereo system, the volume is controlled through an isolating optical link.

Jim Smirniotopoulous

*That's one idea I have never considered. My experience using AC-powered lights to control a photocell has always required filtering out the 60 Hz power-line noise. One easy solution is to build an averaging circuit from a bridge-rectifier, thus driving the lamp with DC....***Steve***

THE VERY BUSY BOX

Dear Steve,

In all of your articles (which I read avidly) I have not seen any

projects directed towards the Heath H8 computer system. I constructed my H8 hoping to learn about computer hardware, but instead found myself only following instructions. I find it very difficult to apply your projects to my system. It would be of great benefit if, in one of your articles, you would include information on interfacing your "house controller" (see "Computerize a Home") to the H8.

Bearing in mind that we H8 owners are basically hardware-oriented, I believe that we would be more likely to construct a project than someone who purchased a system completely assembled. Please consider the H8 in future articles; I am sure that the reception will be well worth the effort.

Ted Benglen

Most computers are equal where interfacing is concerned. If you look closely at the bus signals on your H8 you will notice a striking similarity between their names and the names of signals on the Apple and the Radio Shack TRS-80. The BSR interface (trademarked "Busy Box") requires an I/OWR* strobe (the "*" indicates a negative-true signal), address lines A0 through A7, and power. All address and data bus lines on the H8 use inverted logic levels, so the circuit of figure 1 is necessary to make the system compatible with the TRS-80 attachment shown in the article.

*I generally try to list signal inputs so that experimenters will not be discouraged by a title that says "TRS-80" or "Apple." For simple input and output ports, the signals are often easily accessible and compatible among systems....***Steve***

Figure 1

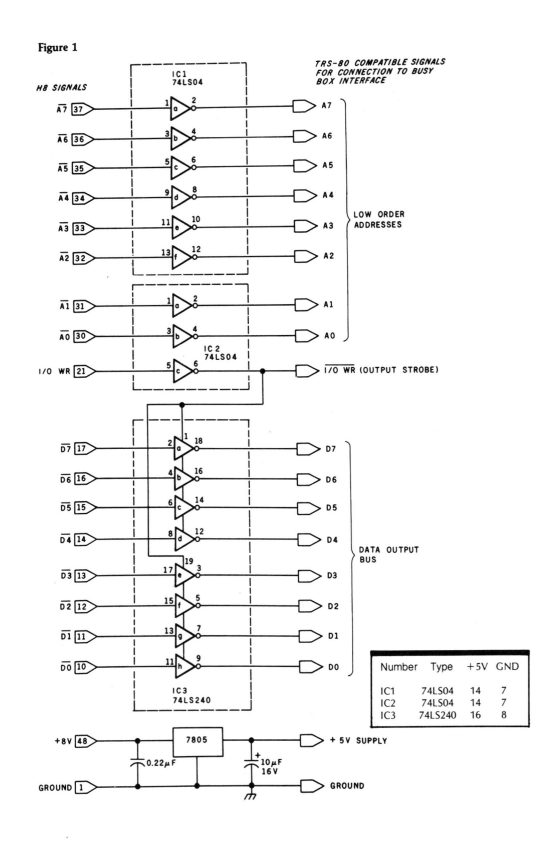

Figure 2

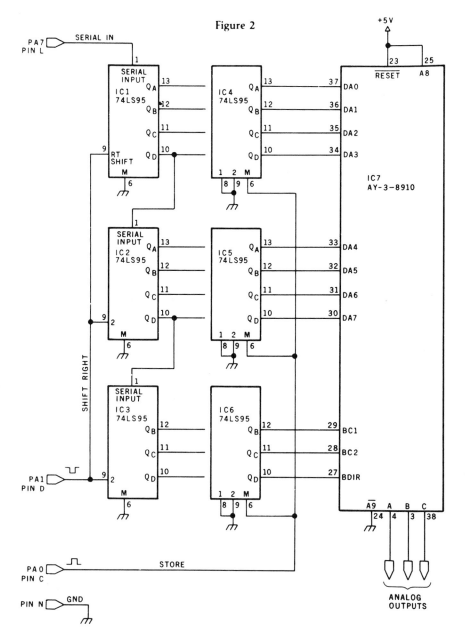

Number	Type	+5V	GND
IC1	74LS95	14	7
IC2	74LS95	14	7
IC3	74LS95	14	7
IC4	74LS95	14	7
IC5	74LS95	14	7
IC6	74LS95	14	7
IC7	AY-3-8910		

TRS-80 is a great noise generator, but I know little of how to deal with the problem. If you can give me any help along these lines, I would appreciate it very much. Thanks.

Robert G Romppel

Radio-frequency interference (RFI) is so pervasive among personal computers and consumer electronic gadgets that the Federal Communications Commission (FCC) has extended the long arm of the law.

As of now, there are various alternatives open to you. First, try plugging the BSR unit into a different wall socket than the TRS-80. The range of the Busy Box is 30 feet, so it doesn't have to be right next to the computer anyway. (Avoid extra long extension cords and use a plug strip for the computer and peripherals.) The noise from the computer is being radiated into the power line; therefore, you want to put as much electrical distance between the TRS-80 and the X-10 as possible. While there may be five wall outlets in an average room, they are rarely all on the same circuit breaker. For the noise to reach an appliance plugged into another circuit loop, it must first travel back to the breaker box. This is a lot of wire and the resulting inductance will diminish some of the interference.

If that doesn't work, next try to kill the noise at the source (the computer) by placing capacitors at the outlet. I suggest using three 0.1 μF 600 V disc ceramic capacitors, one from each side of the AC line connected to a good earth ground and another across the line. Ordinarily, you would also connect

REMOTE CONTROL AT HOME

Dear Steve,

The other day I was thumbing through a *BYTE* magazine and I came across the article you wrote about using the TRS-80 and the BSR X-10 home-control system. I had been working on the same project in my spare time, and I had been using opto-isolators for interfacing; however, your method is well above the idea that I was attempting. Your article was very informative and the accompanying software was excellent. I have since looked up your articles in

other *BYTE*s, and I must say that you never fail to come up with interesting and practical pieces.

I have decided to use your method, and I will shortly be purchasing a "Busy Box" from Micro-Mint in Woodmere, New York.

Whenever I have my TRS-80 up and running, the Sears home-control-unit operation is either marginal or nonexistent. The minute I turn the TRS-80 off, the home-control unit works fine. I assume that the problem is RFI (radio-frequency interference), but I am not quite sure how to cope with the problem. I know the

the computer chassis to ground but this is not advisable on the TRS-80.

To really eliminate line noise, you need a combination of inductance and capacitance. Rather than trying to wind your own coils, it is better for you to buy a commercial noise suppressor. You want one that covers at least a range of 100 kHz to about 200 MHz. They are about $20 and up. One company that lists a few in its catalog is:
Hardside, 6 South St, Milford NH 03055, (800) 258-1790.

If none of this works, then encase the entire thing in copper screening and run it on a battery! ...**Steve**

THE AUTOMATIC APARTMENT

Dear Steve,

I would like to congratulate you on your remote-control article ("Computerize a Home") using the BSR X-10. I have built a unit, and it is now so much a part of my life that I take it for granted. It wakes me up, controls the lights, and guards the apartment in conjunction with a simple burglar alarm.

I have envisioned a system of lighting control that would illuminate any room that I enter, while darkening the one I just left. For this system to work, it must keep track of the number of people in the apartment (if there is more than one person), and it must be able to sense their motion from room to room. Thus, if one person is in the living room and goes to the kitchen, the kitchen light should come on while the living room light should go off. If there is more than one person in the living room, the light should remain on until the last person has left. Of course, manual control should be available, and the system should be able to recognize any sensing errors it may make, and reset itself accordingly.

Obviously, I need a doorway sensor that will detect people passing through, and also detect the direction they are going.

Would you suggest ultrasonic sensors, or would infrared optical sensors be more practical? Could you provide some circuit ideas to help me along?

Jim Porter

I am always glad to hear from someone who takes computer control seriously. Having a computer and automating your apartment makes being "gadget happy" sound almost respectable. In any case, I am familiar with your problem, and I'll try to offer a few circuits that might help.

When I first got involved with security systems, I did a lot of investigation on motion detectors, ultrasonics, and infrared systems. Very few companies offer automatic systems that count people and control lights in rooms. This should give you some indication of what you are getting yourself into.

Two possible methods that come to mind are detecting the motion of people within a room or counting them as they enter and exit.

Motion detectors usually incorporate one of three techniques: infrared, ultrasonic, or microwave. The infrared types are the cheapest. They rely upon changes in ambient light, and the latest designs incorporate an active photosensitive integrated circuit. In fact, Delco Electronics (7 Oakland St, POB 2, Amesbury MA 01913) was offering an under-$30 kit awhile back. In your application, with lights flashing on and off, this may not be a reliable approach.

There are many ultrasonic systems on the market, and they range in price from $50 to $100. My only criticism of them is that they are prone to false alarms and you may find that the harmonics interfere with the BSR system. If you'd like to try placing one across a doorway or diagonally across a room, you could try the circuits shown in figures 1 and 2. These units operate at 23 kHz. Depending upon the sensitivity setting, they will detect most anything passing through the beam. For small rooms,

you won't need much power, so the circuit of figure 1a should suffice. If you need a range of greater than twenty feet, use the higher-power version shown in figure 1b. The receiver for either circuit is shown in figure 2. By the way, the output is TTL (transistor-transistor logic)-compatible. Normally the signal will be a logic 0 (ie: nothing interrupting the beam between the transmitter and receiver); the signal will go to a logic 1 only when someone walks into the room.

The most effective system for detecting motion uses microwave radiation—similar to police radar and operating on the same X-band frequency. In my experience, these are the best by far. They are relatively false-alarm free and very sensitive. I have them installed throughout my home, and I have found their reliability to be exceptional. Unfortunately, they are expensive (in the range of $150 to $400 for domestic installations). A good unit is the Midex 55 made by Solfan (665 Clyde Ave, Mountain View CA 94043). Solfan's more expensive units have contact-closure outputs which would work well in your application.

The final solution to your problem might be to build a people-counter. The circuit in figure 3 (sent to me by William Curlew) might be exactly what you need. It consists of two photodetectors (and two separate light sources) mounted in the doorjamb. Normally the light beam is uninterrupted and the output of the photodetectors is low. As long as there is light on both sensors, the output of IC2b is low. As someone starts through the doorway, one of the sensors goes high, clocking the JK flip-flop into one of two direction states. When the person fully enters the doorway, blocking both the sensors, a trigger pulse is generated and sent to gates 2c and 2d. Depending upon the state of the flip-flop, the clock pulse will be directed to either the count-up or the count-down line of the 4-bit up/down counter, IC5. The counter will increment as people walk into the room and decrement as they walk out. A manual reset is provided to start things out correctly. When the 4 outputs are tied to a parallel input port, your computer can read it as often as necessary to determine how many people are left in the room. Since the counting is done in hardware, timing is not critical. It will accommodate only fifteen people in its present form, so don't have too many guests at your parties! Finally, for absolute certainty, you may want to use it with the ultrasonic circuits previously discussed....**Steve**

REMOTE CONTROL IN EUROPE

Dear Steve,

Please tell me if the X-10 remote-control system by BSR could be operated on 220 V 50 Hz in Europe. I see from the schematic diagrams and various pictures that it is designed to work on 110 V 60 Hz. Do they have a 220 V system? If not, is there any way I could adapt the system to work on 220 V system?

Please tell me where I can buy the set (ie: common console, cordless controller, appliance module, lamp module, in-wall switch module) using an American Express card; maybe from Sears as you said in your article. If so, please let me know the address of Sears, or for that matter, any reliable dealer who accepts American Express. I'll be grateful for the two answers. Next time you are in Europe, drop in and see us. We have a wood stove too, and I hope to connect it to the central heating system.

Rangith Amitirigala
Switzerland

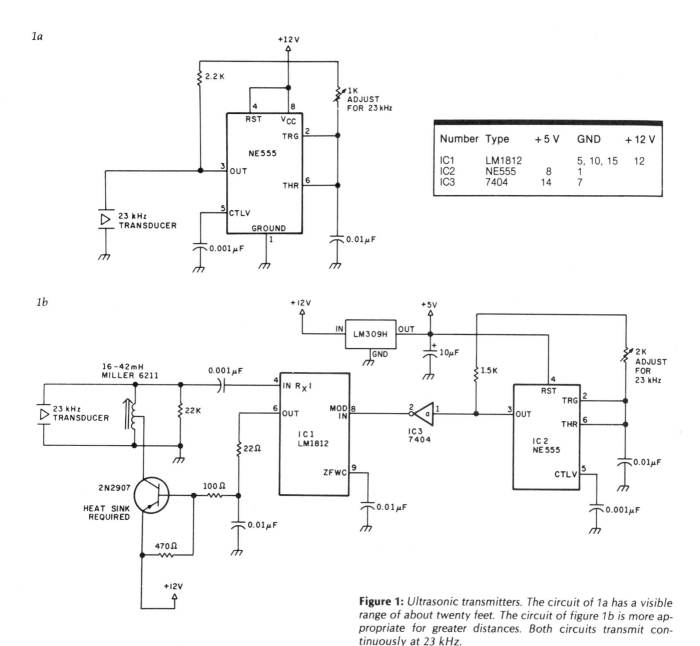

Number	Type	+5 V	GND	+ 12 V
IC1	LM1812		5, 10, 15	12
IC2	NE555	8	1	
IC3	7404	14	7	

Figure 1: *Ultrasonic transmitters. The circuit of 1a has a visible range of about twenty feet. The circuit of figure 1b is more appropriate for greater distances. Both circuits transmit continuously at 23 kHz.*

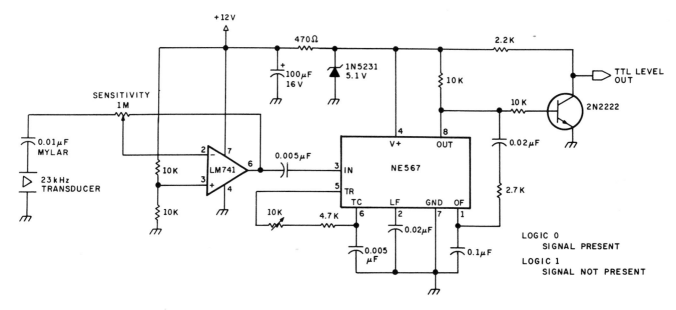

Figure 2: *Ultrasonic receiver. This simple receiver has TTL-compatible outputs, and it will work with either transmitter in figure 1.*

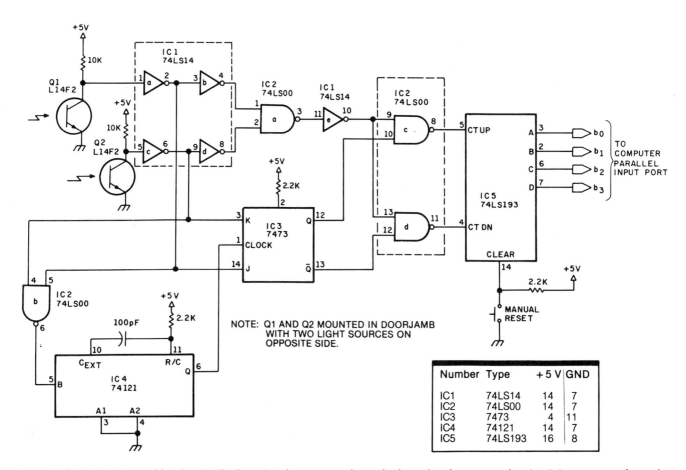

Figure 3: *This circuit is capable of optically detecting the passage of people through a doorway and maintaining a count of people in a room. The photo-transistors sense motion through the doorway and cause the count stored in IC5 (a 4-bit binary counter) to be either incremented or decremented, depending upon the direction of passage.*

Up to this point the X-10 system has been available only in the American version (115 VAC 60 Hz). The custom LSI (large-scale integration) device used in the American units, surprisingly enough, can work on either 50 or 60 Hz. The polarity set on pin 13 of the command-console integrated circuit selects either of the two operating frequencies. These consoles cannot, however, be easily converted from 115 V to 220 V operation without considerable component changes.

A call to BSR (USA) Ltd, in New Jersey, produced some fruitful answers to your question. Even though BSR is working on a European version of the X-10, another company has just announced availability of a 220 V 50 Hz unit. I suggest that you contact this firm for price and delivery. The source is: Busch-Jaeger Elektro GmbH, 5880 Ludenscheid, Freisenberg, Post Fach 1280, West Germany (BRD).

As for Sears Roebuck and Company, it is my understanding that the firm accepts only its own credit card. Rather than worry about which stores will accept your credit card, you may find it easier to go to your local bank (in Switzerland) and arrange for a letter of credit or bank draft when ordering from an American company....**Steve**

The address for BSR is: BSR (USA) Ltd, Rt 303, Blauvelt NY 10913, telephone: (914) 358-6060. There are many stocking distributors for its products including: The Software Exchange, 6 South St, Milford NH 03055.

BSR is an English company, and there may be outlets closer to you than those listed here....**Steve**

ACROSS-THE-SEA FILE

Dear Steve,

I read with great interest your article "Computerize a Home," and I am interested in the BSR X-10 system.

I contacted the Commercial Section of the US Embassy here and also my employer's purchasing agent in New York, but neither could find me the address of the BSR Company. I would appreciate it if you could tell me the manufacturer's address.

Thank you.

Z Lapidot
Rehovot, Israel

A Computer-Controlled Wood Stove

"Come inside, Roger, and get out of the cold." I held my kitchen door ajar as he crossed the front yard towards me. Great clouds of leaves blown by the cold wind furiously encircled him. The landscape was stark and gray, and all weather indicators pointed toward an impending snowstorm.

Roger, a local electrician, had come by to discuss some electrical work I needed done on a new garage I was building. As he stepped through the doorway he remarked, "Sure looks like snow. Have you got enough gas for your Jeep in case you need to plow yourself out of this wilderness?"

Roger's remark reminded me that the terms "picturesque" and "remote" are often synonymous when describing a home in Connecticut. The only place I had been able to buy a house with more than half an acre of land was 25 miles from civilization. And while Roger's controlled, old Yankee humor prevented him from laughing out loud as he spoke, the thought of me, basically a kid from the city, independently plowing my 300 yards of driveway seemed to produce a slow-forming look of amusement.

The Jeep he was referring to was about 20 years old and was used only for plowing. I rather enjoyed the straightfoward task of rearranging snow with it. A certain spirit of excitement came over me each time I stepped into the driver's seat and

asked myself the all-important question posed by every adventurer: "I wonder if this heap will start?"

My neighbor, who shares the chore of plowing, thought I was a sissy when I finally added lights to the Jeep for night driving. Somehow, not seeing the rocks makes hitting them

Photo 1: *The Hydrostove is installed in the corner of the Circuit Cellar. Take note of the two copper pipes coming out the rear of the stove into the wall. The pipes, which are buried behind the wall and above the ceiling, go to the furnace, which is 35 feet away.*

more fun for him. I never did ask him how he had broken the driveshaft the previous year.

I continued my masochistic thoughts of the Jeep. "It should be okay," I said, "but frankly, if it breaks down, I think I'll just hibernate in the cellar for the winter."

Roger still had not taken his coat off as he added, "You might expect to enjoy such an arrangement, but I think you will find that you need outside services more than you think."

"Give me an example."

Roger uncomfortably shrugged his shoulders. Something other than the conversation was bothering him.

"Oil is a good example. You heat with oil, right? How do you propose to fill your oil tank if the truck can't get down the driveway? I'll bet this glass barn you have here almost requires a direct pipeline to the refinery."

I did not exactly relish having my contemporary home called a glass barn but there was some merit to his statement. I retorted, "Who needs..."

Roger interrupted me in mid-statement. "Speaking of heat... what are you running here, a sauna?"

"Take your coat off, Roger. Maybe then you won't be so hot. I'm not so sure you even need both the wool shirt and sweater you have on."

Tossing his coat across to the nearest chair and tugging on his sweater, he continued. "Whenever I

Photo 2: *Logs up to 24 inches in length are placed in a grate which consists of water-filled tubes.*

visit anyone during the winter I presume their house is at 60 degrees like mine. It must be 12 degrees outside, and..." he walked over to the thermostat, "according to this it's 75 degrees in here!"

"I don't usually have it this warm, I was just testing the heating system now that it's computer-controlled."

"What's there to control? Turn the oil burner on longer and it gets hotter."

"Who said anything about an oil burner?"

"Electric heat is even worse!" he quickly added.

"We have oil heat... but it hasn't been on for two days. All I have now is one wood stove."

Roger's momentary blank stare and open mouth were instantly replaced with a look of disbelief. Standing there by the thermostat he quickly scanned the room. With extreme skepticism he replied, "What are you handing me? A twelve-hundred-square-foot room, twelve-foot ceiling, three hundred square feet of glass and seventy-five degrees? I don't see any stove!" Roger walked over to a hot-air duct near one of the windows, stooping down and holding his open palm over the opening he exclaimed, "Wood stove, phooey! There's hot air coming out of this duct. You have the oil burner on!"

"No, Roger. I have a wood stove down in the Circuit Cellar that is plumbed directly into the central heating system."

"A wood stove? In a hot-air heating system?"

"Well actually, Roger, my heating system is both hot water and hot air, and the wood stove heats water. It's called a hydronic wood stove."

"What the heck is a hydronic wood stove?"

Roger was definitely at a loss for words. I put my hand on his shoulder and said, "Think of it as Yankee ingenuity. Come on downstairs and I'll explain how it works."

A Hydronic Wood Stove

A hydronic wood stove is just what the name implies. It is a wood stove that heats water. The particular wood stove that I have is trade-named Hydrostove and it is made by Hydro-Heat Division, Ridgeway Steel, POB 382, Ridgeway PA 15853. Photo 1 shows it installed in the corner of the Circuit Cellar.

The Hydrostove looks like an ordinary wood stove. It is constructed of cast iron and weighs about 400 pounds. The difference between it and a regular wood stove is in the method of heat removal from the burning wood and the ability to channel the energy output into the central heating system.

A regular wood stove produces only radiant energy and is generally a one-room heater unless fans or convection registers are employed to spread the heat around. The surface temperature of such stoves can approach the temperature of the burning wood itself, and great care must be taken to keep combustible material more than 4 feet away.

Typical wood-stove operation is to put in a full load of wood, get it good and hot (warming up the room to around 75° F), and then close the dampers to reduce the heat output. This is the only way to keep the room from becoming unbearably hot. An unfortunate byproduct of this process is that a slow, smoldering fire creates creosote buildup in the chimney. Since only the area directly around the stove is heated, it is likely that an adjacent room will be terribly cold unless fans are used to blow the heat around.

The Hydrostove looks like a regular wood stove, but it operates quite differently. Rather than a solid cast-iron grate, the hydronic stove's firebox is a network of water-filled pipes. These pipes completely encircle the fire, with the burning wood being placed directly on the pipes. Photos 2 and 3 demonstrate this. The inlet and outlet of this water jacket are accessible through two pipe fittings on the rear of the stove. (Since I knew that I wanted a hydronic stove when I built the Circuit Cellar, I had the pipes installed behind the brick wall and

Photo 3: *In operation, the heat from the fire warms the water in the tubes. This is a relatively small fire. The fire box is usually filled.*

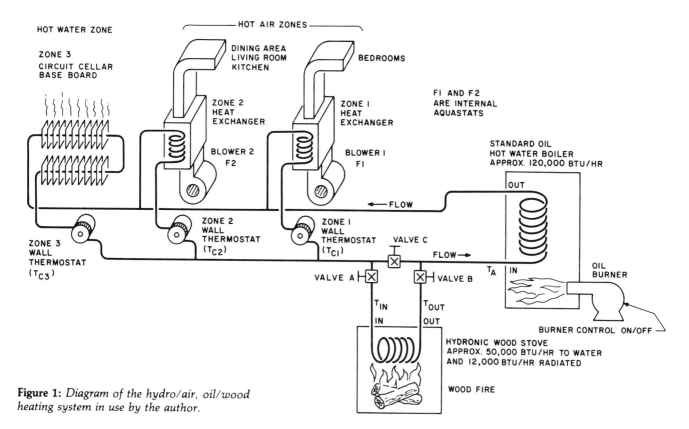

Figure 1: *Diagram of the hydro/air, oil/wood heating system in use by the author.*

through the ceiling. The oil burner is about 35 feet from the Hydrostove.) When a fire is started in the stove, the heat is extracted through the water rather than being radiated directly into the room.

With dry, hard wood, the stove generates about 62,000 BTU per hour (with an additional 12,000 BTU per hour going up the chimney) and is quoted by the manufacturer to be about 85% efficient. I cannot say at this time exactly how much of this is transferred to the water as opposed to how much is radiated. I can only state my experience: with the stove burning at full capacity for 6 hours, the brick wall 1 foot from the stove is on-

ly warm to the touch, and wood can be piled next to the stove (about 2 inches away) with no possibility of ignition. For this same 6-hour period, the Circuit Cellar temperature will never exceed 75° F unless a higher temperature is set on the central heating system thermostat. You would definitely know that it is a hot stove, but anyone inspecting the raging fire inside is usually quite surprised how little heat is felt in comparison to a regular wood stove.

A New England Experiment— First, the Basics

The heating system shown in figure 1 is commonly called a hydro/air

system. It consists of an oil hot-water boiler and hot-air heat distribution. The oil burner heats water, which in turn circulates through a hot-water heat exchanger. A fan blows over the heat exchanger coils and circulates the hot air through the ducts to each room. Such a system combines the even-temperature, residual-heating benefits of a hot-water circulator with the pleasant, humidified, filtered warmth of a hot-air system. A third zone of baseboard heat was added when the Circuit Cellar was built.

Perhaps the best way to start is to explain how an oil-fired hot-water heating system works. Neglect for a moment zones 1 and 2 and the

Note: The heating system in this article is installed in my home and was built to my specifications. I do not intend this as a general construction article, but rather a documented discussion of the elements of the system with emphasis on the controls involved. I must point out that while this article specifically describes a computer-controlled hydro heating system,

general use of a Hydrostove does not require the sophisticated control I have outlined. It is only the unique combination of machinery and an empirically determined operating algorithm that suggests ease of operation through computerization. In truth, the computer's primary value is in the addition of a significant measure of safety rather than the convenience

implied. Through its attachment as a supervisory controller, the computer can more accurately maintain safe operating temperatures and dump excess heat in an over-temp condition. As of the time of this writing, two cords of wood have been burned in the stove testing this complete system and the result has been a safe, satisfying, and reliable operation.

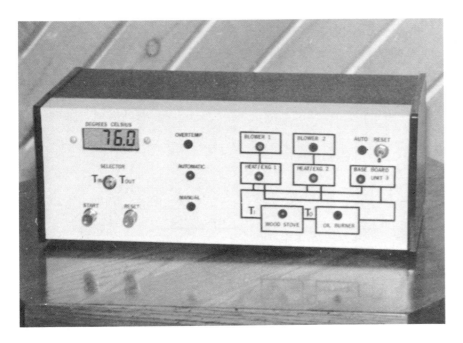

Photo 4: *The computer I/O interface for the heating control system is attractively housed. It includes a display of either the input or output temperature of the Hydrostove and a real-time status display of the circulators and blowers.*

Hydrostove in figure 1. Think of it strictly as the oil burner connected to one circulator pump and the zone 3 baseboard. This is essentially what many homes have. There may be multiple rooms, but only one circulation loop.

Most people think that the thermostat on the wall turns the oil burner on. Actually, this thermostat only controls the on/off operation of the circulator pump; it generally has no direct connection to the burner itself. Operation of the boiler depends upon the temperature of the water flowing into the heating coil section and the temperature setting T_A of the *aquastat* (water conduit thermostat). Water flows from the hot-water boiler to the baseboard and is drawn back through the circulator pump to the boiler again. If the temperature of this water is greater than the aquastat setting, the burner stays off. If however, the temperature is below T_A, the burner turns on, adding heat until the water in the loop reaches T_A. Usually T_A has a wide hysteresis; the high and low limit of variation is separated by about 20° F. For most boilers the low setting is 160° F, and the high is 180° F. The hysteresis reduces the frequency of oil burner starts.

To get heat in a room, you turn up the wall thermostat, which starts the pump. As the water moves through the baseboard, it loses heat to the room. The water is then reheated by the oil burner.

Now, consider the addition of the Hydrostove as shown in figure 1. Any water circulating through zone 3 will necessarily pass through the coils of the stove if valves A and B are opened and C is closed. This circulation in itself does nothing to the operation of the heating system. If, however, you build a fire in the Hydrostove as in photo 3, heat is added to the water returning from the baseboard and flowing into the boiler. If the fire is large enough, the temperature of the water flowing out of the Hydrostove is greater than T_A, the oil burner never turns on, and the house will effectively be heated by the Hydrostove.

There are a few other considerations. Unlike the oil burner which can be selectively turned on when heat is needed, once the wood stove is on, it runs for quite a while and the heat must be *continuously* removed; otherwise, the water in the pipes will turn to steam. Pressure-relief valves will keep the system from exploding, but who wants a steam bath in their living room? In a single-zone system, the circulator pump must remain on until the fire is out. In a gravity-feed system, the pump must stay on until the fire is lowered to the point where

the water stays below the boiling point and can effectively be radiated by the heating loop.

Consider the Hydrostove as a continuous source of heat. If the Hydrostove is cranked up to produce 40,000 BTU per hour, then 40,000 BTU per hour must somehow be removed. The task of heat dumping is much easier on a multi-zone system. Take for example, three zones with capacities of 40,000 BTU, 30,000 BTU, and 20,000 BTU, respectively. Whether or not a room thermostat is calling for heat, you must turn on either the pump for zone 1, or the pumps for both zones 2 and 3. Consider the case when zones 2 and 3 are used as heat dumps. If the zone 1 thermostat were to trip suddenly, the control system would have to make a choice. It could add zone 1 to the pool and share 40,000 BTU among three zones or immediately drop zones 2 and 3 off the line and send everything to zone 1 until it reaches its thermostat setting again. While the previous choice can easily be made, load sharing is an interesting consideration. It is much easier to switch zones on and off while performing load sharing than to try to directly control the heat output of the wood fire to any degree.

An additional complication occurs when using heat exchangers. Heat exchangers cannot effectively transfer heat unless the blower is on. The fans in these units are thermostatically controlled. When the water flowing through the exchanger reaches a set temperature, the fan turns on, extracting heat. There is considerable delay and overshooting in the operation of these units. While the *average* hourly heat transfer of a heat exchanger might be 40,000 BTU, it may be 10,000 BTU with the blower off and 50,000 BTU with it on. In a quick heat-dump situation, it is sometimes necessary to override the blower thermostat and force the blower on to maintain stable conditions throughout the rest of the system.

Using a Hydrostove

How much you benefit from the addition of a hydronic wood stove depends quite heavily on the rest of your heating system. Above all, it must be capable of taking the full heat output of the wood fire. This can be

62,000 BTU per hour. Since my oil burner is rated at 120,000 BTU per hour, and I had added the third zone of baseboard to the Circuit Cellar, I concluded that the connection would be quite safe.

My usual method of manual operation is to use the stove only on very cold days and to build as large a fire as possible. It is initially started with both dampers open, but once the fire is going strong the flue damper is closed to reduce the amount of heat going up the chimney. At the time the fire is started, the zone 1 circulator pump is turned on continuously with a switch, overriding the motor-start relay. This keeps some water flowing through the stove at all times. Zones 2 and 3 are normally left in their "heat on demand" thermostat-controlled mode. If the Circuit Cellar cooled down and its circulator pump kicked in, it would be drawing heat from the stove along with zone 1.

Our house is large, but given my method of use, no single heating zone can sustain the full output of the Hydrostove for long periods of time. Generally, the water temperature will be between 75° and 90° C. To maintain a 20° to 22° C (68° to 72° F) temperature through the house on a very cold day, I have to keep the fire box continually filled. This means filling the stove with wood every 3 to 4 hours. (Before you choke and compare it to 12 hours for a regular airtight stove, remember that I am talking about heating a whole house). After a few hours of use, even in this large house, the temperature in the rooms in zones 2 and 3 will reach the wall-thermostat set points, no longer continuously demanding heat from the stove. This leaves all the heat going to zone 1.

Soon, the temperature of the water coming out of the wood stove starts to climb above the safe high limit of 88° C (measured 35 feet away at the furnace). When the indicator hits around 98° C, a loud noise can be heard in the pipes because the higher temperature water nearest the hot coals within the stove is turning to steam. Unless you want the safety valve to blow, filling the room with steam, you have to override the automatic settings of either or both of the thermostats of zones 2 and 3 to get rid of some of the excess heat. It may also be necessary to manually turn on the heat-exchanger blowers

for zones 1 and 2 for the reasons I previously outlined.

This occurrence is rare, and I generally have about 10 minutes to react to the situation and throw all the manual switches required. After using the system and determining that this is a potential problem, I installed a digital temperature indicator that allowed me to monitor the system as I worked at my desk in the Circuit Cellar. When I saw that the temperature was going above 88° C, I would throw the manual override on the zone 2 circulator pump. If the temperature did not drop, I would continue with the other heat dumping methods. It has never gotten to a point where these maneuvers prove insufficient or where the fire has to be put out. Experience has shown that zone 2's volume of 15,000 cubic feet provides a terrific sink for excess heat. Normal use in this mode barely raises its temperature more than 2° F above its nominal 66° F thermostat setting. (I do not want to leave you with the impression that there is one 90° F room in the house.)

I have not described anything thus far that specifically requires computerization other than for convenience. The real reason is a rather insignificant detail that is discovered only after actually using the stove. A Hydrostove definitely saves oil, as

stated. When its output temperature is greater than the aquastat set point, the oil burner does not come on. The problem arises during startup and shutdown when the stove output temperature (T_{out}) is less than the setting of the aquastat (T_A). The circulator pump has to remain on while there is a fire but, because the circulation loop is running and returning at less than T_A, the oil burner keeps coming on. *Catch 22!!*

If this were a matter of 10 minutes or so, it would not be so bad; but shutdown to the point where the circulator pump can be turned off can take several hours. The alternative is to cut the power to the oil burner when the wood stove is on and restore it when the fire is out. This is what I initially did, until I was staying up all night to shut the stove off. The alternative was to wake up to a very cold house, turn the oil burner on, and have to wait a half-hour to take a hot shower.

Aside from taking in a tenant who would watch the Hydrostove temperature in exchange for room and board, the only reasonable alternative was a more intelligent control system. With the proper sensors, a device could monitor the heat output of the stove ($T_{out} - T_{in}$), and when it dropped to a predetermined safe point, automatically restore power to

Control Outputs Signal	Type	Function
TC_1	Contact closure 200 mA	Circulator pump - zone 1
TC_2	Contact closure 200mA	Circulator pump - zone 2
TC_3	Contact closure 200 mA	Circulator pump - zone 3
F_1	Solid state relay 5 A 220 VAC	Heat exchanger blower zone 1
F_2	Solid state relay 5 A 220 VAC	Heat exchanger blower zone 2
XFER	Contact closure 5 A 115 VAC	Oil burner power

Inputs Status	Level	Function	
TC_1	TTL		
TC_2	TTL	0 pump off	
TC_3	TTL	1 pump on	
F_1	TTL	0 blower off	
F_2	TTL	1 blower on	
T_{in}	Analog		
		range 40° to 240° F	
T_{out}	Analog	4° to 115° C	

Table 1: *Computer I/O lines used with the heating control system.*

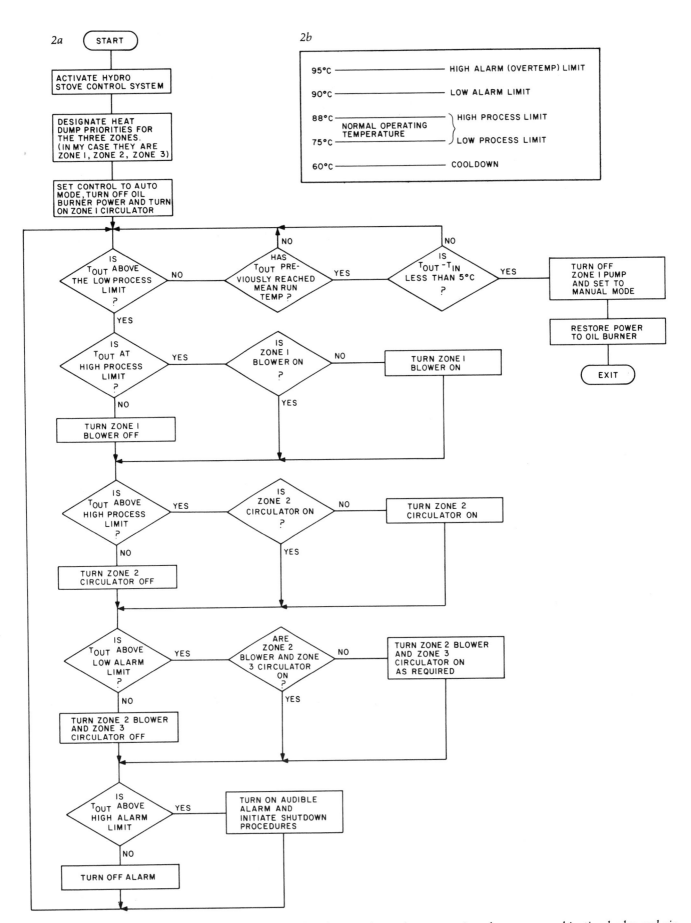

Figure 2: *(a) Logic flow for the automatic distribution of Hydrostove heat when output in a three-zone combination hydro and air heating system. (b) Points of importance in the operating temperature range of the system. The actual set points for process and alarm limits depend upon placement of temperature sensors and may vary a few degrees.*

In BYTE, algorithmic flow is assumed to proceed down and to the right unless an arrowhead is present to indicate otherwise.

Type of wood Hardwoods	Pounds per cubic foot	BTU per cord	Equivalent gallons (gallons per cord) fuel oil
1. White Ash	37.5	23,037,000	165
2. Cherry	31.0	19,043,920	136
3. Hickory	45.0	27,644,400	198
4. Maple (red)	33.5	20,579,720	147
5. Oak (chestnut)	41.0	25,187,120	180
6. Walnut	34.5	21,194,040	151
7. Willow	24.0	14,743,680	105
Softwoods			
8. Douglas Fir	30.0	18,429,600	132
9. Ponderosa Pine	25.0	15,358,000	110
10. White Spruce	25.0	15,358,000	110

Table 2: *Comparison of wood heat values for various species of wood available in North America and their equivalent in gallons of fuel oil per cord of wood. (These estimates are generally accepted by industry.)*

the burner and shut off the circulator-pump override.

Computer-Controlled Heating System

My heating system is not technically a computer-controlled wood stove. It is rather a system designed specifically to efficiently distribute the heat from a wood stove, to safely dump excess heat in an effective manner, and most importantly, to restore the entire system to its standard configuration when the fire is out. I am merely outlining one application of the many that are conceivable when the heating system has been connected to a computer. Complete energy management is a possibility; or, at the very least, total energy output can be closely monitored and recorded. I am working on these areas, but for now, the topic is control.

Virtually any personal computer can suffice as the controller. The logic is straightforward and relatively uncomplicated. It is outlined in the flowchart shown in figure 2. Proper control of the three zones and the Hydrostove requires a special interface to connect the computer to the various blowers and pumps. Table 1 is a list of the signals in question.

The control outputs from the computer are, in essence, all contact closures, whether it be through mechanical or solid-state relays. The use of relays provides electrical isolation between the computer and the heating system. It further prevents potentially dangerous loops between 115- and 220-VAC powered components.

The three zone thermostats are low-voltage AC circuits that can be directly controlled through a reed relay, as shown in figure 3. The relay contacts are connected in parallel with the thermostat. With the thermostat contacts open, a logic 1 control signal closes the relay and provides an alternate current path to pull in the pump-start relay. By monitoring the voltage across the relay contacts, it is possible to directly monitor the activity of the circulator pump and determine its operational status at any time. If the contacts are open, current flows through the optoisolator light-emitting diode (LED), producing a logic 0 status at the output. When closed, no current flows and the logic value is 1. My application required only the ability to turn on a pump which may not already be running. However, to accommodate complete functional control of the pumps, the thermostat can be disconnected as shown.

The interface to the heat exchanger blowers, shown in figure 4, is similar. This time, however, a solid-state 7 A, 220 VAC relay is used. The power to the blower is 5 A, 220 V rather than low voltage AC as before. A 7 A solid-state relay was chosen because of its size and low cost.

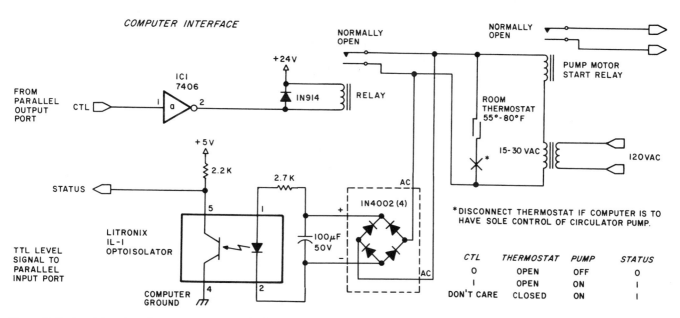

Figure 3: *Isolated interface for computer control of a typical oil-fired, hot-water, 1/4-horsepower circulator pump. The 7406 open-collector inverter (IC1) requries a 5 V supply to pin 14 and a ground connection to pin 7.*

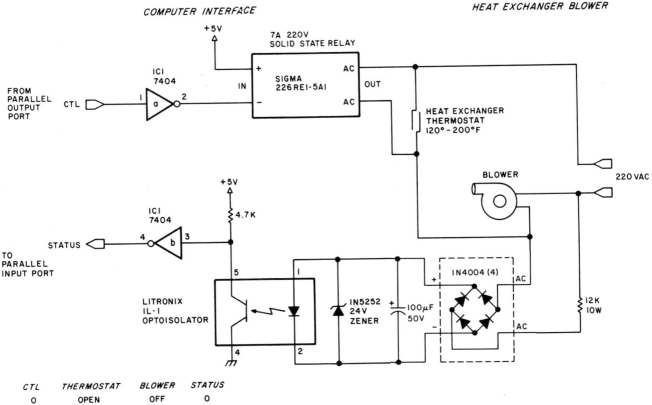

CTL	THERMOSTAT	BLOWER	STATUS
0	OPEN	OFF	0
1	OPEN	ON	1
DON'T CARE	CLOSED	ON	1

Figure 4: *Isolated interface for computer control of a heat-exchanger blower fan. The 7404 hex inverter requires a 5 V power supply to pin 14 and a ground connection to pin 7.*

Monitoring the activity of the blowers is accomplished simply by checking the voltage across the motor. The 220 V present when the motor is on is reduced and rectified to run an optoisolator as before. With voltage present, the status output is high (logic 1).

Finally, the computer must be able to monitor the output temperature of the Hydrostove. This signal is an analog voltage that is proportional to temperature. Various sensors such as thermistors or thermocouples could be used, but a more practical device is a temperature sensor device such as the LM334 from National Semiconductor. When configured as in figure 5, the output of IC1 (monitored at V_{in}) is 10 mV per degree Celsius. It may have a nominal offset of something like 2.5 V, but if the temperature rises 10° C the output will go up 100 mV. ICs 2 and 3 provide gain and offset adjustment and are configured to prohibit accidental negative excursion of the output if the temperature sensor goes open circuit. The result is a circuit that converts a change in temperature to a change in voltage. By adjusting the gain and offset, 0° C can be an output of 0 V and 100° C can be 1 or 10 V. A Fahrenheit scale can be just as easily calibrated by setting a different gain and offset.

To read this signal, the computer must have an analog-to-digital converter interface. This can be either a true successive-approximation analog-to-digital converter as in figure 6, or the discrete set-point level detector of figure 7. The choice

Photo 5: *An internal view of the I/O controller containing relays, optoisolators, and analog interface components.*

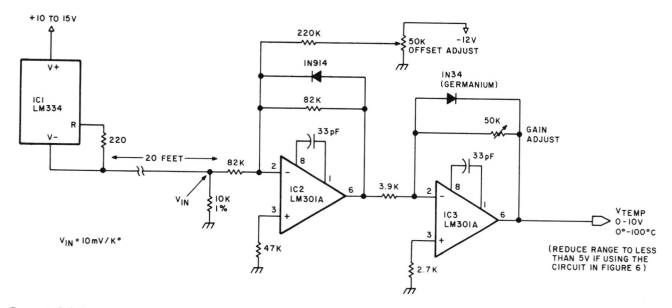

Figure 5: *Solid-state temperature sensor. The range of the output voltage (V_{TEMP}) is reduced to less than 5 V if the circuit in figure 6 is used.*

depends upon whether you need to know the exact temperature or just significant set points.

If data acquisition is the dominant consideration, then consider the circuit of figure 6. IC8 is an 8-channel, 8-bit analog-to-digital converter that is bus-compatible with most microprocessors. Figure 8 outlines its internal structure. As configured, it is attached to function as ports F8 through FF, with port F8 corresponding to input channel 0, and port FF corresponding to channel 7. The volt-age on channel 0 is read by initiating an output to port F8. This causes the address of 000 to be stored and the conversion process started. After about 100 microseconds, the time necessary for conversion, the channel analog value can be obtained by reading an input from port F8. A similar procedure is used to set and read the other channels.

If you are interested strictly in control, then the circuit of figure 7 is much simpler to use. If a 0 to 10 V input represents a range of 0 to 100° C and there are eight comparators, each could be set to trigger 12.5° C higher than the preceding one. A better approach is to arrange the majority of set points to cover the control and alarm range rather than to cover insignificant temperature ranges. For example, bit b_0 could be set to trigger at 60° C. It is not necessary to care much about temperatures below that point. The range of prime interest is from about 75° C to 95° C. Dedicating 5 set points within this range, another perhaps between 60° C and 75° C and a final overtemp indicator at 98° C should prove more than adequate.

My system uses a combination of both interfaces, using set points for control inputs and a true analog-to-digital converter to determine actual heat output from the stove.

A further enhancement is a visual display indicating the real-time status of the system components and a readout of the actual temperature. The prototype controller is shown in photos 4 and 5. It serves as the interface between the heating system and the computer, and contains most of the electronics described in this article as well as other enhancements not discussed at this time. While all the control decisions are actually made by the computer, the display gives me the added satisfaction that everything

Photo 6: *To effectively use this control device, it is important to have accurate temperature measurements. The LM334 temperature sensor is easily attached to the Hydrostove return pipe by wrapping Teflon plumbing tape around it.*

Number	Type	+5 V	GND	−12 V	+15 V
IC1	LM334	see figure 5			
IC2	LM301A			4	7
IC3	LM301A			4	7
IC4	LM339		12		3
IC5	LM339		12		3
IC6	REF-01	see figure 7			
IC7	LM301		4		7
IC8	ADC0808	see figure 6			
IC9	74LS30	14	7		
IC10	74LS02	14	7		
IC11	7400	14	7		

Table 3: *Power and ground connections for the integrated circuits that are used in the circuits of figures 5, 6, and 7.*

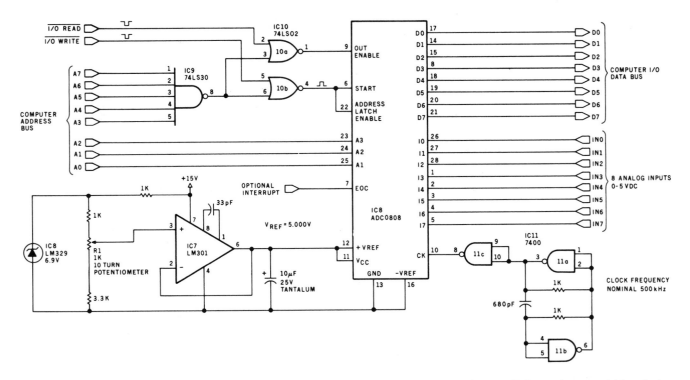

Figure 6: *An 8-channel, 8-bit analog-to-digital (A/D) converter using a National Semiconductor ADC0808 data acquisition device.*

is working correctly.

Back to Roger's Visit

Roger studied the stove very carefully. He was surprised at the simplicity of the idea of supplemental wood heat, but somewhat aghast at the overall complexity of the entire system. The concept of computer control did not concern him in the least but I sensed that my apparent independence from OPEC fostered a little competitive jealousy.

"What about wood? You still will have to get wood for the stove," Roger commented, pointing out a possible serious limitation.

"I'm surprised you didn't notice when you drove in. There are eight cords of wood piled outside. I don't expect to use them all this winter. Wood, unlike oil, is one of those things you can easily stockpile if you have enough storage space."

"Oh yeah, I did notice a few piles beside the driveway."

Roger was perplexed. He had obviously begun to believe the petroleum company propaganda. The thought of missing an oil delivery meant total destruction of civilization as far as he was concerned. But he just could not believe that the addition of a wood stove meant independence. Suddenly he smiled as he thought of a sobering reality that I might have overlooked.

"You have to keep the circulator pumps running when the stove is going, right? And if the fire is real hot you may in fact need the blowers on as well?"

"Sure, why?"

Roger had found the Achilles' heel of my heating system. The Hydro-stove as I had it configured needed power to run all the pumps and blowers. The actual heat might come from a wood fire, but distribution of the heat throughout the house depended upon the local electric utility. Roger quickly commented, "What happens if the power goes out?"

"Well, I suppose I should be concerned, but I'll have four or five minutes to react."

Roger laughed. "React to what? Living in a steam bath?"

"Perhaps I should show you. Follow me." I led Roger out of the cellar into the garage. In one corner was a large mechanical contraption, part of which was a two-cylinder engine. Pipes and wires came to it from different directions, all converging at a central control box adjacent to the motor. Without explaining the intricate details involved with this permanent installation (the heating system was enough for Roger this time) I said, "If the utility power goes out, I throw the emergency transfer switch and start my 5-kilowatt generator. It's large enough to run the whole house and then some."

Hesitating, then striking out with one last effort, "You still need gasoline and that doesn't look like a very big tank."

"Sorry Roger, I thought of that, too. This particular unit runs on both gasoline and propane. There's a 100-gallon propane tank outside the garage just for the generator."

"I give up!"

It is just as well that he did. Eventually he would notice the trench going across the driveway from the house to the new garage. When he is installing the wiring for it I hope he doesn't ask why I am running insulated copper pipes underground across to the garage. ∎

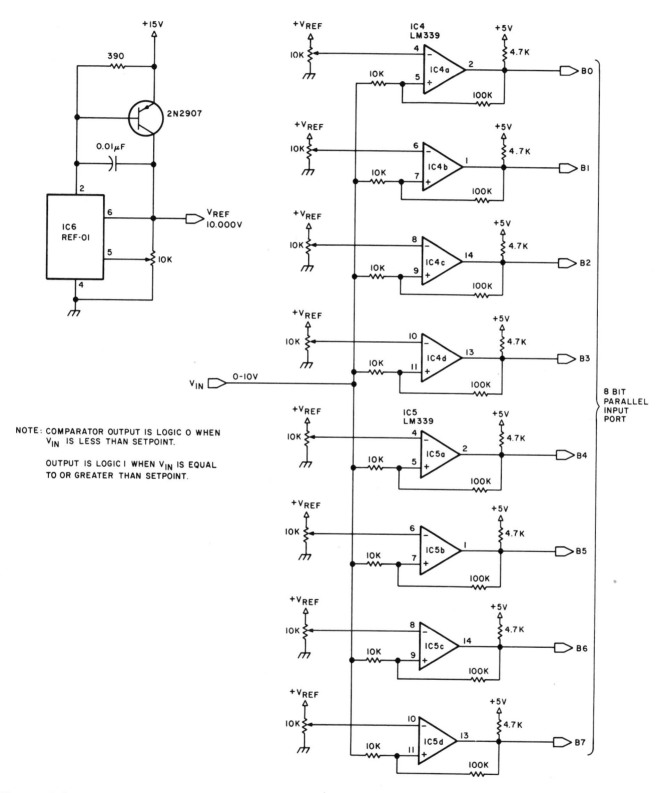

Figure 7: *A discrete set-point-level detector. This method is cheaper than the method shown in figure 6 and can be used only when it is necessary to detect a small number of temperature ranges. The eight comparators on the right-hand side of the figure are wired to have their outputs go from logical 0 to logical 1 when a certain temperature (determined by the position of the 10 K potentiometer) is exceeded. The status of the eight bits can be used to determine what range of temperature the interface is currently in. The voltage reference integrated circuit REF-01 (IC1) may be obtained from Precision Monolithics, 1500 Space Park Dr, Santa Clara CA 95050.*

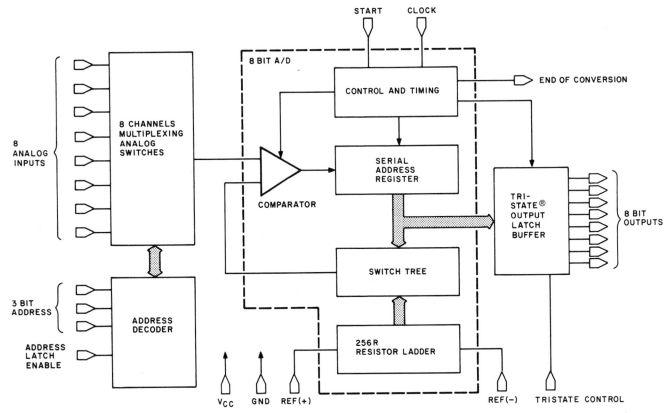

Figure 8: *National Semiconductor's ADC0808/0809 CMOS data acquisition system. The 8-bit converter uses successive approximations. The device interface is to most 8-bit microprocessors.*

LEVELS TO BITS

Dear Steve,

I have been shopping around for the analog-to-digital (A/D) converter integrated circuit that you used in your woodstove interface, but it does not seem to be readily available.

C W Vuaun

*I try to avoid specifying components that are not commonly available. While I obtain parts through industrial distributors rather than surplus outlets, I check the latter often to see what is available. In the case of the ADC0808, the time-lag is greater than I expected. However, in the meantime there is a sixteen-channel version, the ADC0816CCN, which is the same in every respect (except that it has twice as many channels). It is available from Digi-Key Corporation, POB 677, Thief River Falls MN 56701. Their toll-free phone is (800) 346-5144. Call or write them for the current price....**Steve***

A HOT TIP

Dear Steve,

The solid-state sensor you described for your wood stove in "A Computer-Controlled Wood Stove," is very interesting. I have constructed the circuit, but I am having trouble calibrating the device for a range of −18 to +100°C.

Ron Goodmaster

The circuit you refer to can be calibrated in a number of ways. There is an offset and gain adjustment included for this purpose.

*In normal practice, say for a range of 0° C to 100° C, we would adjust for offset so that the output was 0 V with the temperature probe in an ice bath and adjust the gain so that the output is 1.00 V when it is placed in boiling water. To have it actually read −18° as −0.18 V you will have to modify the circuit slightly. Presently, the 50 k offset-adjustment potentiometer is connected between +12 V and ground. By connecting it instead between +12 V and −12 V you can impress a negative current flow into IC2 such that it has a negative offset. The gain of the circuit will now have to be adjusted for a 118-degree span instead of 100 degrees. The trick is that to accurately calibrate the unit, you should have a −18° C standard when you set the low end. Substituting a voltage source for the LM334 will only give you a relative calibration, but it may be all you need....**Steve***

Ease into 16-Bit Computing:

Get 16-Bit Performance from an 8-Bit Computer

Stopping for coffee at the local doughnut shop has become a morning ritual. I am quite capable of making coffee at home, but I am not what you would call a "morning person." Even though I have culinary talents that include the preparation of eggs Benedict and strawberry crepes, it had better be evening when you request them around our house.

This morning started out like any other. I pulled my car into the doughnut shop's parking lot only after carefully examining all the potential hazards. I carefully avoided the broken glass, the beat-up 1962 Chevy and the large black van with a "Tax the Rich!" bumper sticker.

After entering the shop, I sat down and spread my reading material, the latest issue of *BYTE*, on the counter. As my coffee and bran muffin were delivered, I could not help but overhear the conversation of two other people at the counter.

"Dave, have you been reading any of the magazines lately? It looks like everyone is going 16-bit crazy."

"I've read a lot of descriptive articles, but I suppose it'll take a while before we see any real hardware."

"Actually, I'm a little hesitant to just jump on the bandwagon. My 8085 works just fine."

"I know what you mean, Ed. The Z80 system I built from scratch is still cranking along. I'd like to do something with the 16-bit chips, but I sure don't want to throw out my 8-bit system."

"What about building a small system to experiment with? Didn't I see an article a few months ago on a single-board 8086?"

"Yeah, I remember. It was in *BYTE*. Wasn't it written by that guy who lives around here someplace, in his cellar or something?"

Upon hearing that last statement, I nearly choked on my muffin. I thought it would be prudent to remain anonymous until I learned whether or not they enjoyed the article. I carefully closed the magazine and placed it face down on the counter.

One way to ease yourself into the world of 16-bit computers is with the Intel 8088. This microprocessor is an 8086 on the inside with an 8-bit data bus on the outside.

"Maybe, but anyway, the article wasn't too bad," said Ed. I'm sure they didn't hear the sigh of relief from across the counter. Then he continued, "But it just seemed like a larger computer than I have time to build. It's obviously oriented toward guys who don't have any other development system. I'd prefer a minimal hardware configuration to start with. If I want large programs, I'll run a macroassembler on my 8085 system, write the object code into an EPROM, and then plug it into the test board."

"Eliminating all the keys and displays will help, but how small a computer can we end up with and still be 16-bit? You'll need 16-bit address and data buses, and what's 1 K words of memory—four chips? All the EPROMs I know are 8-bit output. That means at least two of them."

"Wait a minute," said Ed. "I didn't say I had all the answers. The minimal configuration may be twenty chips, but isn't this closer to something we could afford to experiment with?"

This was the perfect opportunity to express my point of view concerning the things that I write and consult about. "Excuse me," I said. "I couldn't help but overhear your conversation. Had you considered using an 8088?"

The two young men looked up at me, paused, and harmonized, "An 80 what?"

"I know a little about microprocessors. Have you considered using an Intel 8088?"

"Is it 16-bit?" asked Dave.

"Well, yes and no," I replied. "It uses an 8-bit data bus, but, internally it's an 8086. Essentially it's an 8-bit chip that's completely 8086-software-compatible."

Should they listen to this doughnut and coffee philosopher? "That sounds tremendous, but won't it still require quite a few chips to make an operational computer?"

I sensed that this was a good time for my exit. Staying any longer would involve my designing a computer for them on the back of a napkin. Ordinarily I probably would have stayed, but I had just completed a similar task in my latest article, so I decided to let them wait a few more weeks. I rose to leave, carefully rolling up the copy of *BYTE*, cover page inside, and stopped behind them on my way out. "My recollection is that while four chips is a possibility, a five-chip computer is quite a reality. I've even seen how a BASIC interpreter could be written to run on it. In case you're interested, the next issue of *BYTE* has an article all about it."

I excused myself to attend an important meeting. As I opened the door I heard, "Thanks, I'll look forward to reading it." They watched me intently as I drove out. I could only speculate on their final conversation.

The 16-Bit Generation

The exciting items in microcomputing these days are the 16-bit microprocessors made by companies such as Intel (the 8086), Zilog (the Z8000) and Motorola (the M68000). All of these devices, although they differ in internal architectures, commonly claim to have compressed the power of a minicomputer within a single chip of silicon. Most notably are the 16-bit data bus and increased addressing space. A 20-bit address can directly address a megabyte of memory.

There seems to be little doubt in the minds of microcomputer-system designers that the 16-bit processors are the wave of the future. Already some major manufacturers are designing the new processors into intelligent terminals, word-processing systems, and other equipment. The day when this revolution within a revolution will affect the personal and small-business computer marketplace is not too far away.

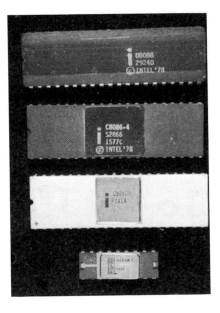

Photo 1: *An exhibit of advancing microprocessor technology. Here are four integrated circuits produced by Intel Corporation. From bottom to top, we have the 8008, the first 8-bit general-purpose microprocessor; the 8080A, one of the breed of 8-bit devices that helped ignite the microcomputing boom; the 8086, the advanced 16-bit processor; and the 8088, the subject of this article—a component that contains 16-bit computing capability in a package that can communicate with the outside world through an 8-bit data bus.*

But if it is obvious that the 16-bit machines will be the trend of future product technology, it is equally obvious that it is relatively difficult for the designer to make a leap from the 8-bit world of the 8080, Z80, 6800 and 6502 to the emerging 16-bit world. The 16-bit instruction sets are more complex. The 8086, for instance, has a repertoire of some 133 instructions, as compared to seventy-eight for the 8080. Simply because of the larger range of memory that can be addressed and because of address segmentation, addressing of memory is more advanced. Also, the register set is more complicated, and the types of operands with which the processor can work are more extensive.

As complex as the 8086 or any other 16-bit microprocessor is from a software viewpoint, it is in the design of hardware circuits to work with the 16-bit processors where the real complexities arise. Peripheral interfaces and existing hardware systems are generally based on an 8-bit data bus. When your whole design is built

to make efficient use of an 8-bit data bus, converting to a 16-bit architecture is not a simple matter of replacing the processor. This incompatibility dictates substantial design changes to take advantage of the new 16-bit microprocessor.

A Gradual Approach to 16-Bit Computing

There is an alternative to converting abruptly to 16-bit architecture. Look at photo 1 and observe the Intel 8088 microprocessor. This device uses an 8-bit data bus, so all of your present hardware components will work with it from the standpoint of getting information between the processor and the peripheral-support devices or memory, but the 8088 features a common internal architecture and complete software compatibility with the 16-bit 8086 processor.

As a result, the 8088 provides an excellent way for designers, engineers, hobbyists, and students to ease into the world of 16-bit computing. Its 8-bit-compatible bus structure makes it the logical choice for upgrading 6800, 6502, Z80 and 8080 designs to 16-bit capability without alteration of existing 8-bit hardware.

The 8088 can be used in projects such as a low-cost system that employs multiplexed peripherals such as the 8155, 8755A and 8185. Or, fully expanded, it forms a system that allows a full megabyte of address space and compatibility with the 8086 family of coprocessors and multiprocessors.

This two-part article is designed to give you a glimpse of the 8088. In Part 1, I shall attempt to familiarize you with the instruction set of the 8088 and the hardware of a microcomputer that is made from an 8088 and only four other integrated circuits. The power of this five-chip circuit will be emphasized by illustrating, among other examples, how it can be configured to support a multi-user Tiny BASIC.

Architecture of the 8088

Anyone comparing the internal architectures of the 8088 and the 8086 processors will realize that they are identical. Even though I have previously discussed the 8086, a brief explanation of this architecture is necessary since the capabilities of our five-chip computer depend directly upon it. However, if you wish to read a more detailed description, you should refer to the previous Circuit Cellar article, "The Intel 8086," page 102.

A diagram of the internal structure of the 8088 is shown in figure 1. The 8088 contains two logical "units," the bus-interface unit (BIU) and the execution unit (EU), and a 4-byte instruction queue.

The *execution unit* is where the actual processing of data takes place inside the 8088. It is here that the familiar arithmetic and logic unit (ALU) is located, along with the registers used to manipulate data, store intermediate results, and keep track of the stack. The execution unit accepts instructions that have been fetched by the bus-interface unit, processes the instructions, and returns operand addresses to the bus-interface unit. The EU also receives memory operands through the bus-interface unit, processes the operands, and then passes them back to the bus-interface unit for storage in memory.

The role of the *bus-interface unit* is to maximize bus-bandwidth utilization (that is, to speed things up by making sure that the bus is used to its full capacity). The bus-interface unit carries out this assignment in two basic ways:

- by fetching instructions *before* they are needed by the execution unit, storing them in the instruction queue
- by taking care of all operand fetch and store operations, address relocation, and bus control. (These actions of the bus-interface unit leave the execution unit free to concentrate on processing data and carrying out instructions.)

Figure 2 summarizes the 8088 register set. The shaded registers are the 8080 register subset, that is, the registers that are common to the 8088 and its 8-bit predecessors.

The *general registers*, also called the HL group because they can be subdivided into *High* and *Low* bytes, include the accumulator (AX), base (BX), count (CX) and data (DX) registers. The AX register can be addressed as a 16-bit register, AX, or the high-order byte can be addressed as the register AH and the low-order

Photo 2: *An exhibit of advancing memory technology. The single black integrated circuit at the center can replace the entire board of components. The center component is the Intel 8185 1 K-byte static programmable memory. The board is a 1 K-byte memory board from a Scelbi 8B microcomputer system, which used the 8008 microprocessor (circa 1975).*

Photo 3: *Using the 8088 and other components of kindred technology, it is possible to build a functional microcomputer system with only five integrated circuits. Part 2 of this article will present more detailed information about this system.*

byte as AL. The same holds true of the other three general registers (BX, CX, and DX).

Another group of registers is the *pointer and index* (or P and I) group. This set contains the stack pointer (SP), base pointer (BP), source index (SI), and destination index (DI)

registers. Generally speaking, these registers hold offset addresses used for addressing within a segment of memory. They can also participate, along with the general register group, in arithmetic and logical operations of the 8088.

The 8088 uses memory segmenta-

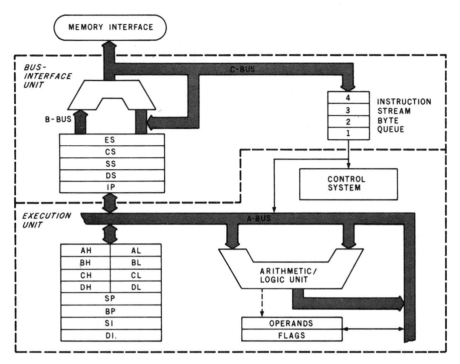

Figure 1: *Diagram showing internal operational principles of the 8088 microprocessor. The 8088 (and the 8086) use a pipelined architecture that increases performance by overlapping instruction execution with memory-fetch operations. The 8088 can directly execute any 8086 software.*

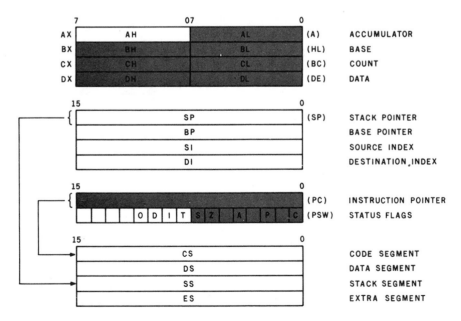

Figure 2: *The 8088 contains fourteen 16-bit registers. The shaded registers are those common to the 8088 and the 8080.*

tion to address this large memory space efficiently. At any one time, the 8088 can deal with memory as a set of four 64 K-byte segments. The total memory is organized as a linear array of 1,048,576 bytes, addressed as hexadecimal 00000 to hexadecimal FFFFF. The 8088 creates a 20-bit address by combining a 16-bit offset and a segment boundary value stored in one of the segment registers. Figure 3 demonstrates how this works.

Each of the 16-bit-segment registers, the code segment (CS) register, the stack segment (SS) register, the data segment (DS) register, and the extra data segment (ES) register, contains a value that is added to a 16-bit offset address, forming a 20-bit address. The memory is thus divided into a maximum of four 64 K-byte segments that are active at any single time. The *code segment* of memory is where instructions are stored, the *stack segment* of memory is where the pushdown stack is located, the *data segment* is where data to be operated on is found in memory, and the *extra segment* is an additional 64 K-byte data area.

When fetching an instruction from memory, the location accessed is given by a 20-bit address that is the sum of two numbers. The first number is the value of the 16-bit instruction pointer. The second number is a 20-bit value that is the 16-bit code-segment register with four low-order zero bits appended. This forms the 20-bit address required to specify any location in the megabyte-sized address space.

In the case of a memory-reference operation for a transfer of data, the absolute memory address referenced by a given memory-access instruction is calculated by adding the given 16-bit address to the base address. The base address is given by the contents of the data-segment or extra-segment register and is followed by four low-order zero bits.

In the case of a stack operation, the memory location referenced is similarly offset from the value contained in the stack-segment register.

The 8088 has both relative and absolute branch instructions. When all branch instructions within a given segment of memory are specified in relation to the instruction pointer and the program segment does not modify the value of the code-segment register, the program segment can be relocated dynamically anywhere within the megabyte address space. A program is relocated in the 8088 simply by moving the code, updating the value of the code-segment register, and resuming execution.

Small System Applications

The 8088 can be used in a broad range of applications, from systems requiring use of a minimum number of components to systems requiring maximum performance. The component-count-sensitive applications include point-of-sale terminals and simple controllers, which require that system cost be kept low, but need substantial processing power. A big reason for this design flexibility is the ability of the 8088 to operate in a minimum-hardware mode.

The minimum-mode, multiplexed configuration, as shown in figure 4, is an effective way of building a powerful system around the 8088, while using the smallest number of parts. The processor is connected in the minimum mode by wiring its Mn/Mx pin in the high-logic state (at V_{cc} potential). The multiplexed bus is directly compatible with the Intel 8085A-family peripheral components (8155, 8355, 8755A, and the new 8185).

A four-chip system can be designed using the following components: an 8088 microprocessor; an 8284 clock generator; an 8155 memory, input/ output (I/O), and timer device; and an 8755A EPROM and I/O device. A fifth component, the 8185, is a simple addition to the system and provides an extra 1 K bytes of user memory.

In the minimum-mode configuration, the 8088 provides all necessary bus-control signals, including $\overline{RD}$, $\overline{WR}$, IO/$\overline{M}$, and ALE. It further provides HOLD and HLDA (hold-acknowledge) signals to allow direct-memory-access (DMA) data transfer, INT and $\overline{INTA}$ to interface the 8259A interrupt controller, and $\overline{DEN}$ and DT/$\overline{R}$ to control transceivers on the data bus.

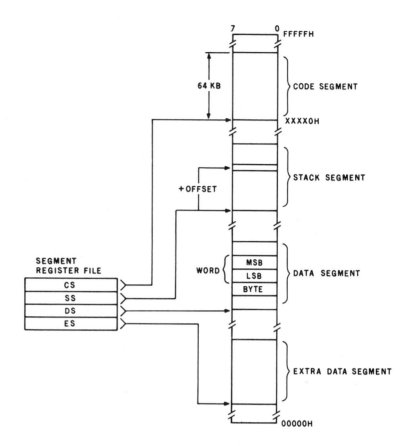

Figure 3: *Memory organization. The 8088 uses a memory-segmentation technique to address up to 1,048,576 bytes (1 M byte) of memory. The user can use attributes of the memory-addressing system to dynamically relocate a program anywhere within the entire address space.*

The power of the 8088 can be extended in large-system applications by wiring it into the maximum-mode configuration. However, a discussion of maximum-mode features is beyond the scope of this article.

The 8088 Instruction Set

A complete discussion of the 8088's instruction set is also beyond the scope of this article. Rather than attempt it, I shall concentrate on some specific features of the 8088 instruction set that facilitate the specific application discussed in Part 2 of this article. These features include extended-arithmetic instructions, direct use of ASCII-encoded data, multiprocessing features, string-manipulation instructions, and table-translating aids. The 8088 instruction set includes single-instruction multiplication and division instructions, along with five different types of addition and seven types of sub-

traction operations.

These multiply and divide instructions greatly facilitate "number crunching." This numerical ability saves much time in such applications as data sampling, signal processing, and scientific calculation. Not only are fewer machine instructions needed to perform a given task, with corresponding savings in memory usage and execution time, but the versatility of the instructions and the ability of the 8088 to deal with several types of data remove the usual necessity of handling messy conversions from one type of data representation to another and back again.

Two program listings demonstrate the saving of effort. Listing 1 gives the 8088 code for the skeleton of a subroutine that accepts data from a specified input port and calculates a running average of the values entered. The same subroutine section coded for the older 8080 micro-

Listing 1: *An example of the efficiency of the 8088 and 8086 instruction set. This short routine accepts input of five values from an input port, and then calculates and sends a running-average value to an output port. Compare this listing with listing 2.*

```
            XOR     BX, BX        ;CLR BX
            MOV     CX, 5         ; Set loop counter

Average     INC     BL            ;Increment data counter
            IN      AL, Port #    ;Input data
            ADD     BH, AL        ;Update running total
            MOV     AL, BH
            DIV     BL            ;Divide running total by
                                  ;data counter.

            OUT     Port #, AL    ;Output running average.

            LOOP    Average       ;Return unless fifth pass
                                  ;is completed.

            HLT
```

Listing 2: *A routine that performs the same task as the routine given in listing 1. This code, however, was written for the older 8080 processor. As you can see, it is longer and more tedious to write.*

```
            MVI     H,00          ;Clear H register
            MVI     E,00          ;Clear E register

Average     INR     E             ;Increment data counter

            MOV     C, H
            IN      A, Port       ;Input data
            ADD     H             ;Add data to running total

Divide      XRA     A             ;Clear accumulator
            MOV     B, A          ;Clear B register
            MOV     L, A          ;Clear L register
            MVI     C, 80         ;Initialize bit counter

Loop        MOV     A, C          ;Shift B and C as
            RAL                   ;a 16-bit unit—
            MOV     C, A          ;one bit left
            MOV     A, B
            RAL
            MOV     B, A

            CMP     E             ;Compare data
            JC      Next          ;counter (divisor) with
                                  ;dividend; if divisor is larger,
                                  ;bypass subtract.
            SUB     E             ;Divisor is smaller; subtract.
            MOV     B, A
            MOV     A, D          ;Set current bit of
            ORA     L             ;L to 1
            MOV     L, A

Next        MOV     A, D          ;Shift D right and check carry
            RRC
            JNC     Loop          ;If no carry, return for next bit.

            MOV     A, L          ;Outport running average
            OUT     Output

            MVI     A, 05         ;Return unless fifth pass is
            CMP     E             ;completed.
            JNZ     Average

            HLT
```

processor is shown in listing 2.

Direct Use of ASCII and Decimal Data

The direct use of unpacked binary-coded decimal (BCD) or ASCII-encoded data in a microcomputer has a number of obvious advantages. Since many I/O devices present data to the processor in American Standard Code for Information Interchange (ASCII) format and expect responses in the same format, microcomputer-system designers have for years faced the necessity of putting their input and output through a translation process (usually involving a table look-up operation) before processing the input or responding with output.

With the 8088's instruction set, such manipulation is no longer necessary. All four mathematical instruction types (add, subtract, multiply, and divide) provide for ASCII adjustment of the accumulator contents by a single instruction. This feature is obviously of great use in everyday microprocessor applications. Equally interesting (and useful) are the two instructions that adjust the results of addition and subtraction to *packed* decimal form.

Table-Translating Aid

Despite the availability of single instructions to convert accumulator contents from one type of data representation to another, it may still be necessary from time to time to translate data by means of the traditional look-up table. This might, for example, be necessary if the data is being received or transmitted in EBCDIC (Extended Binary-Coded-Decimal Interchange Code) rather than in ASCII form.

The XLAT (ie: translate) instruction allows the user to define a 256-byte table of correspondence and then to reference any point in the table very easily. The base address of the table is placed in the BX register and the index (ie: table position) is stored in the accumulator. Then the single instruction code XLAT is used to refer to the proper point in the table, pick out the translation, and store the result in the accumulator.

This is useful particularly when data that has been entered from a

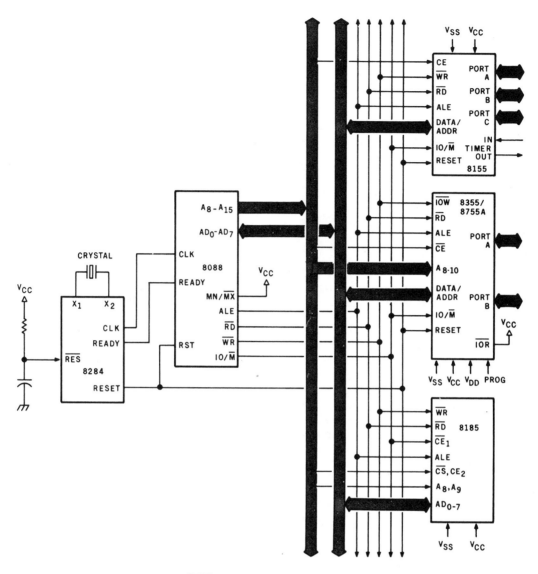

Figure 4: *When used in the minimum mode (MN/ $\overline{MX}$ line held high), the 8088 interfaces directly with the multiplexed address and data components in the 8085A-support family to form a functional microcomputer system using only five integrated circuits. Detailed information concerning this circuit will be given in Part 2.*

port comes into the accumulator for disposition or transfer. If you are dealing with a stream of incoming characters in EBCDIC format, for example, the translation proceeds as follows: you begin by storing the beginning memory address of your 256-byte translation table in the BX register. If you set up the table so that the base address of the table corresponds to an incoming EBCDIC value of 00, the next address to an incoming value of 01, etc, all you must do is simply accept a byte of data and execute the XLAT instruction.

This simple procedure lets us obtain the correct translation of that byte into the proper format for han-

dling by the 8088 or some other processor. A MOV instruction will then store the result of translation until it is needed; the translation process can then be repeated with the next incoming byte. Setting up the necessary instruction sequence requires one instruction: a MOV to the BX register of the base address of the table. The loop for handling the translation requires only three basic instructions: the input instruction, XLAT, and MOV.

String-Manipulation Instructions

Since typical computer applications often deal with strings of characters consisting of letters,

numbers, and special symbols, easy-to-use string-manipulation instructions are a welcome enhancement to 8-bit processors. The 8088 addresses this need by providing five powerful primitive string operators that may be preceded by a single-byte repetition prefix.

For a byte-for-byte or word-for-word comparison of two data strings (as you might use in verifying the accuracy of data loaded into memory from a mass-storage device, for example), the 8088 offers the CMPS instruction. This also allows termination of a program segment upon occurrence of a predetermined equality or inequality condition, as well as automatic incrementing or decrementing.

You can scan through a string of data for an occurrence or for an absence of occurrence of a specific string or character by using the SCAS instruction. This operation subtracts the byte or word operand in memory (or elsewhere) from the accumulator and changes the logic state of the flags; it does not, however, return a result. Again, decrementing or incrementing is automatic.

The STOS instruction allows you to fill a string of arbitrary length with a single value (eg: a string of zeros or nulls for a floppy disk initialization routine), once more with automatic incrementing or decrementing of a predetermined count.

Putting Some Things Together

Let's take a quick look at a small but powerful example that employs both the string manipulation and the XLAT instructions to solve a very practical problem.

You are designing an input routine that must translate a buffer filled with EBCDIC characters into ASCII form, continuing the transfer until one of several possible EBCDIC characters is received. The transferred ASCII string should be terminated with an EOT (end-of-transmission, hexadecimal value 04) character. Assume that the buffer starts at hexadecimal memory location FFFE, the table to translate the EBCDIC form to ASCII begins at hexadecimal location 0100 and the CX register is to contain a value giving the length of the buffer

Listing 3: *A segment of 8088 code that translates characters from Extended Binary-Coded-Decimal Interchange Code (EBCDIC) to American Standard Code for Information Interchange (ASCII) form. The 8088 instructions for manipulating and translating strings of characters are put to good use.*

```
        MOV     SI, FFFE      ; Source index register contains start of EBCDIC Buffer
        MOV     BX, 0100      ; B register points to translate table
        MOV     DI, ASCBUF    ; Destination index points to ASCII buffer
        MOV     CX, 528       ; C register contains length of buffer
        CLD
        JCXZ    EMPTY         ; Skip if input buffer empty
NEXT:   LODS    EBOBUF        ; Get next EBCDIC character
        XLAT    TABLE         ; Translate to ASCII
        STOS    ASCBUF        ; Transfer ASCII character to buffer
        CMP     AL, EOT       ; Test for EOT character
        LOOPNE  NEXT          ; Continue if no EOT received (CX decrements first)
        .
        .
        .
EMPTY:  (Program continues)
```

containing EBCDIC characters. The buffer may, of course, be empty.

The small 8088 program segment shown in listing 3 accomplishes this task in a small number of instructions and handles a great deal of overhead work with little effort or concern on the part of the system designer and programmer.

By now you should have an understanding of the power of the 8088 microprocessor. Even in a minimal-mode, five-component circuit, our little computer will have the following attributes:

- 5 MHz 8088 8-bit processor (completely 8086 software-compatible)
- 1280 bytes of static user memory
- 2048 bytes of erasable, programmable read-only memory (EPROM)

- 38 parallel I/O lines
- a 14-bit counter/timer
- power-on reset and nonmaskable interrupt.

In Part 2, we will deal with some key features of the 8088 which make it particularly suited to multiprocessing situations. We will investigate the operating system of a multi-user, Tiny BASIC language system on our minimal-configuration computer.∎

These figures are provided through the courtesy of Intel Corporation.

16-BIT SYSTEMS

Dear Steve,

I was fascinated by your article "Ease into 16-Bit Computing," and I am impressed by the added computing power available through the new 16-bit microprocessors; but I am interested in building a complete system, not just a computer. This practically requires that the processor board be S-100 compatible. How long will it be until these specifications are met, and will it be worth the wait? Which one will have the edge in performance under personal computing conditions?

Lynne Poderson

*I have written articles about the Intel 8086 and 8088 processors, and I generally like what I hear regarding the Motorola 68000. I try to refrain from making too many claims for hardware that I have not personally checked out. There are 8086 S-100 boards presently available, but it will be awhile before any 68000 or Zilog Z8000 units are available. Industrial users are acquiring most of the parts, so prices are currently high. My opinion is that these third-generation microprocessors will all perform in the same league when used in personal computers. The true measure of efficiency will be a function of well-written operating systems and high-level language implementations....*Steve

Dear Steve,

Last spring I acquired an 8086 "University Kit" from Intel, with the intention of breadboarding a system. Several tinkers dissuaded me, saying that the high frequencies involved would not be suitable to my intended medium. I decided to shelve the idea until a reasonably-priced S-100 processor board became available. The boards that are presently available are all in the $500 range. Would I be better off buying an 8088 and building your five-chip system described in the March and April "Circuit Cellars," or are cheaper boards going to be available soon?

Tom Boerjan

I disagree with the person who told you not to prototype an 8086 system because of the high frequencies involved. The frequencies are no higher than those used in the 8080. The key, of course, is good layout and proper construction techniques. You can still try an 8088, but a bird in the hand...

The 8088 is still new and University Kits did not exist when I wrote the article. An S-100 board for either processor will be expensive. Only time and increased chip production will lower the cost.

*For a printed-circuit board for the 8088 system described in my article, write to John Bell Engineering, POB 338, Redwood City CA 94064. They should be able to provide the necessary components. For application notes on the 8088 processor, write directly to Intel Corporation at 3065 Bowers Ave, Santa Clara CA 95051....*Steve

Ease Into 16-Bit Computing

Part 2: Examining a Small Multi-User System

In computer club meetings, in software-development groups, and among hardware designers, the terms multiprogramming, multiprocessing, and multitasking are often heard. Now that we have a few years of experience in microprocessing, the prefix *multi* has become prevalent. I define multi as an indication of the ability of a system to seemingly process more than one function at a time.

Multiprogramming, as I refer to it, is a form of program execution that allows more than one user to access the resources of a computer system at (apparently) the same time. Rather than denoting the execution of multiple programs simultaneously, which requires the use of more than one processor, multiprogramming implies a division of a single processor's time and resources. A computer executes commands faster than any single human user can enter data or instructions. A user in such a situation may never realize that there are other users connected to the same computer.

Because the input and output are being performed by the operator at human speed (which is extremely slow relative to the speed of the microprocessor), most of the processor's time in a single-user system is spent waiting for the operator to enter information, or for an output device to display the information being sent by the processor. The ratio of time the computer spends in useful activity to time the computer spends waiting is very small. Multiprogramming takes advantage of this relatively large amount of wait time by using it to execute a request from one of the other concurrent users. Of course, as the number of users on the system increases, the operator response time

---

Surprisingly little hardware is required to support a multi-user system running Tiny BASIC.

(ie: the amount of time it takes for the computer to respond to a specific request from an operator) will become longer and longer until it reaches some unacceptable limit. In order to maximize the number of users that may use the system concurrently with acceptable response time, the operating system may be tailored to a particular type of application.

Your first question may be, "How much hardware is required to support a multi-user system running a high-level language such as BASIC?" The answer: surprisingly little. Because of the 16-bit processing features of the Intel 8088, which I outlined last month, a multi-user operating system can be provided with a computer consisting of as few as five integrated circuits.

It is beyond the scope of this article to discuss and list the entire assembly code of the Tiny BASIC system written for the 8088. The assembly listing of the 2 K-byte interpreter is thirty-one pages long.

Readers who are interested in using the 8088 for a similar application are advised to contact the manufacturer directly. Intel is publishing an application note describing a small (seven integrated circuits) multi-user Tiny BASIC system that uses the 8088. There was discussion at the time of this writing (January 1980) that a printed-circuit board of the expanded circuit would be available for sale as well.

For this information contact:
Tom Cantrell
Marketing Communications
Intel Corporation
3065 Bowers Ave
Santa Clara CA 95051

Minimum System Hardware

The five integrated circuits required to build a workable system include the 8088 microprocessor; the 8284 clock generator; the 8155 memory, input/output (I/O), and timer device; and the 8185 erasable programmable read-only memory

(EPROM) and I/O device, all from Intel Corporation.

The 8088, residing in a 40-pin package, executes the complete 16-bit instruction set of the 8086 microprocessor, while communicating over an 8-bit data bus. The 8088 was discussed in Part 1.

The 8155, shown in figure 1a, is also in a standard 40-pin dual in-line package (DIP). It provides 256 8-bit words of static memory and is powered by a single +5 V power supply. Since it is static memory, no refresh circuitry is required.

In addition to the memory and a programmable timer, the 8155 also provides two programmable 8-bit I/O ports and one 6-bit programmable I/O port. The high-order bit of port B is chosen as the serial input line for one of the two user terminals, and the low-order bit of port A is used as the serial output line for the same terminal.

Figure 1b presents the internal block diagram of the 8755A. The 8755A combines EPROM and I/O functions. The EPROM contains the system software; the I/O ports serve the second user's terminal.

The last major part in the system is the 8185, which contains 1 K bytes of static memory. (See figure 1c.) It is used by the system as the major block of memory allocated for program storage.

All of these integrated circuits are specifically designed to work with the multiplexed address and data buses. Hence, there is no need to have any outside latches to provide address signals for their operation. Address latching for each device is provided internally.

All of the integrated circuits used in this design are directly compatible with the 5 MHz signal which is generated by the 8284 clock generator (figure 1d); however, the 8155 timer/counter appears to work better if driven by the 2.5 MHz signal that is output on the PCLK line of the 8284.

Figure 2 is a diagram that demonstrates the flow of data in the 5-chip system, as well as the addresses of the memory and the I/O ports. To allow for service to multiple simultaneous users, the *timer-in* line in the 8155 is wired to the PCLK line in the 8284. Also, the *timer-out* line is tied to the nonmaskable interrupt (NMI) line of the 8088 microprocessor.

Developing the Operating System

There are many problems associated with writing a BASIC interpreter for such a limited system. Approaches taken on large computers are not necessarily applicable. Tiny BASIC is usually written to work with one user taking up all of the resources of the system. In this case, the problem is to share the resources and allow more than one user (in this case, two users) to access the processor and the memory without interfering with any other user in the process.

Allowing for the input and the out-

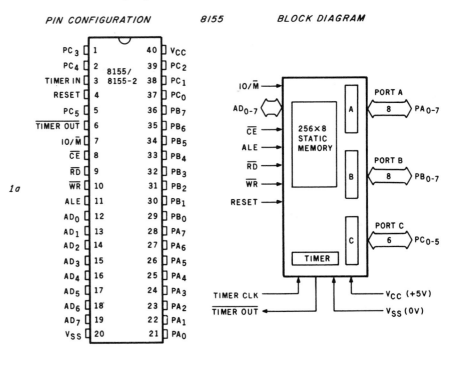

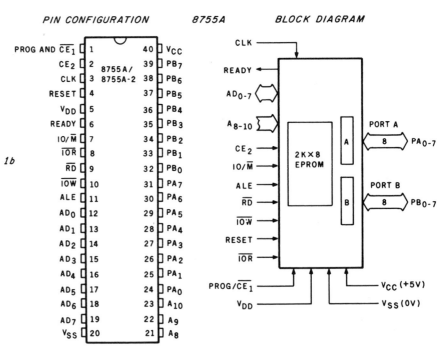

Figure 1: *Integrated circuits that perform support functions for the 8088 in the minimum-configuration system discussed in this article.*
(1a) The 8155 static memory, I/O, and timer device.
(1b) The 8755A EPROM and I/O device.
(1c) The 8185 1 K-byte static memory part.
(1d) The 8284 clock generator/driver device.

put of the different users is easy, since the 8155 and the 8755 both provide two 8-bit I/O ports. All that is needed is to use one of the two for input and the other for output. One bit of data is shifted in or out at each interrupt from the timer.

The data rate for communication with the user terminals is obtained by using the programmable timer in the 8155 as a data-rate generator. The 14-bit binary counter is preset during the initialization routine of the system. Once set, the counter continuously counts up and generates an interrupt signal when it reaches the specified value.

The value set in the counter determines the data-transfer rate. In this system the counter value is contained in the EPROM, and is therefore not easily changed. The data rate must be chosen and the counter value computed before programming the EPROM.

Dividing the memory between users is an easy task. All that is needed is to assign each user specific areas to be used for program space, buffers, and stacks. This does limit the size of the programs that may be entered by each user, but from an operating-system viewpoint, the assignment of space is an easy task. A memory map is outlined in figure 3.

The problem of memory allocation in this situation is getting the processor to differentiate between users, buffers, and programs. Since there are 2 K bytes of EPROM to contain all of the system programs, it would be easy for the operating system to

PIN CONFIGURATION   8185   BLOCK DIAGRAM

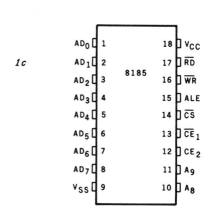

1c

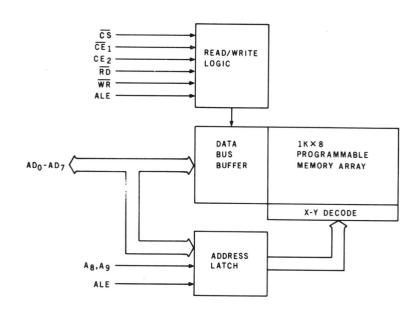

PIN CONFIGURATION   8284   BLOCK DIAGRAM

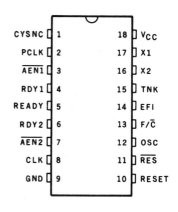

1d

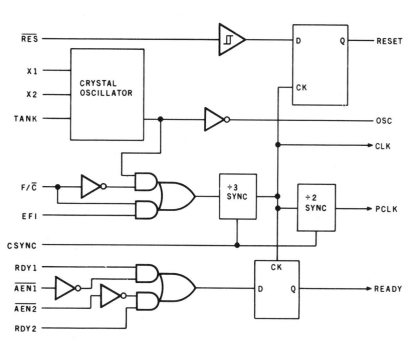

use all of the memory space in just initializing the two user terminals. An easy, efficient method of differentiating the two users is required.

Another consideration in the interest of total system efficiency is the allocation of more execution time to one of the users if the other user has his job executing some kind of I/O wait loop. Normally, the processor will switch the current user-task being executed each time it receives an interrupt from the timer. This way, each user-task will receive an equal amount of execution time on the system.

However, while the system is waiting for a user to enter commands or while it is sending information to the terminal, it has no productive task to perform for that user. If both users are in an I/O mode, as at system-startup time, then the processor enters a wait loop, waiting for the interrupts from the timer. This way, as much as possible, the processor will split time with both users effectively.

Solving the Problems

The biggest concern, differentiation between the two users and their respective buffers and programs, was the easiest to solve with the 8088 microprocessor. This processor, like the 8086, addresses all memory locations using one of four *segment registers*.

All of the jumps and subroutine calls within a program are made relative to the current position of the execute properly if the segment registers are set up correctly.

It is also this segmenting feature that allows us to write the BASIC interpreter in such a way as to address the buffers and programs instruction pointer. Hence, the jumps and calls are not specific to the memory segment where a given section of program code is placed. The code can be moved from place to place within memory, and will still belonging to one user in a relative mode, and to modify the actual memory area being accessed by just changing the segment registers to point to the area containing the specific user-task we currently want to work with.

Specifically, the 256 bytes of user memory in the 8155 are divided into two areas, one for each user, to provide the required stack buffers. The 1 K bytes of user memory in the 8185 are divided into four areas for each user. User 1's stack buffer goes from hexadecimal locations 10 to 7F. User 2's stack buffer goes from hexadecimal locations 90 to FF.

Corresponding areas in the two stack buffers are separated by hexadecimal 80 bytes. Each of the buffers in the program buffer area of memory (contained in the 8185) is separated by hexadecimal 200 bytes. These memory areas are shown in figure 3.

When the microprocessor needs to access a given area in memory, the effective address of the memory that is to be accessed is computed by multiplying the appropriate segment register by 16, and then adding the

Photo 1: *This 8088 system fits in the palm of your hand and uses only five integrated circuits. It contains enough read-only and programmable memory, and sufficient peripheral interfaces, to support two 300 bps terminals, running a Tiny BASIC interpreter on each.*

Photo 2: *Side-by-side comparison of the 8755 EPROM, top, and the older 2708 EPROM. The difference in package size is due to the presence of I/O ports on the 8755. The 8755 requires a single 5-V power supply and contains 2 K bytes of memory; the 2708 requires three different power supply voltages (+5 V, +12 V, and −12 V) and contains 1 K bytes of EPROM.*

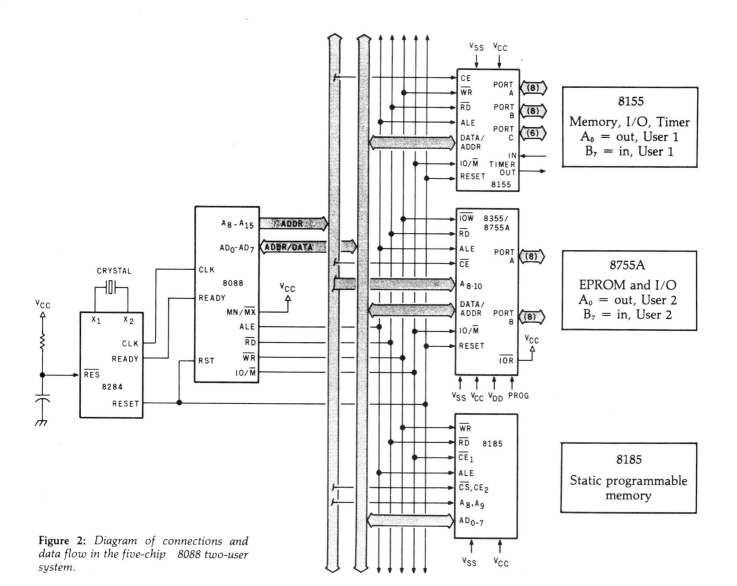

Figure 2: *Diagram of connections and data flow in the five-chip 8088 two-user system.*

result to the position within the segment.

For example, if the processor was instructed to load the byte at hexadecimal location 154 within the segment, and the data-segment register contained the hexadecimal value 14, the resulting *effective* address is computed as:

$$14_{16} \times 10_{16} = 140_{16}$$
(data segment value times 16)
$$\underline{+ \ 154_{16}}$$
$$294_{16}$$
(location within the segment)

Therefore, if I want the processor to access user 1's pushdown-stack buffers, I set the stack-segment register equal to 0. When I access the stack buffer, which is located from hexadecimal addresses 10 to 7F, the effective address computed will still be hexadecimal 10 to 7F.

Device	Type	Hexadecimal Address
8155	programmable	00 thru FF
8185	programmable	1000 thru 13FF
8755A	EPROM	0F800 thru 0FFFF

Table 1: *Memory addressing in the 8088 system.*

Device	Function	Hexadecimal Address
8155	Command-Status Register	00
	Port A	01
	Port B	02
	Port C	03
	Timer (low)	04
	Timer (high)	05
8755A	Port A	F000
	Port B	F001
	Data-Direction Register A	F002
	Data-Direction Register B	F003

Table 2: *I/O addressing in the 8088 system.*

If I want to access user 2's stack, I set the stack-segment register to a value of 8. When the processor computes the effective address, it will multiply the stack-segment value by 16 and add the product to the location within the segment. This means that user 2's stack buffer will be correctly addressed in hexadecimal locations 90 thru FF while allowing the program to use the same address values used to access user 1's stack.

The program buffers are handled in essentially the same way. For user 1, the data-segment register and extra-segment register are set to 0, and the program is written to address the buffers as hexadecimal addresses 1000 to 11FF. When I want to access user 2's program, I load the segment registers with the hexadecimal value 20. When the processor computes the effective address, it will come up with hexadecimal addresses 1200 thru 13FF, which is what I want.

Since the interpreter itself does not modify values in the segment registers, the interpreter never knows which user-task it is currently working on, but it does not care. With the proper loading of the segment registers by the *operating system*, the correct buffer of the current user will be used.

Using this feature, the 8088 processor can work for several users, switching between them by manipulating only the segment registers. Because of memory limitations, the maximum practical number of users on the system described here is only two. However, the programs could just as easily serve three or four users as two users.

Software Modifications

There are two other software routines that must be specifically modified to handle multiprocessing. The initiating sequence of code that is executed when a restart signal is received must be changed. Also, an interrupt handler for the nonmaskable interrupt generated by the timer of the 8155 must be added.

When the microprocessor is reset, the initiating routine initializes all the I/O ports and sends out the initial stop-bit signal to the terminals. It also sets up user 2's stack area so that the processor will begin execution at the START routine when it is through processing user 1. After setting the correct data-transfer rate for the user terminals, the initiating routine jumps to START for user 1. The initiating routine is required so that the registers, buffer areas, and the stacks will be set up properly for each user before any other processing begins.

Once normal processing has begun, the routine that handles the timer-out interrupt switches the user-tasks on each interrupt cycle and determines when it is time to input and output information to the terminals.

The timer-out routine is shown in flowchart form in figure 4. This interrupt-handling routine is the key to getting the other software to process multiple users.

When called in response to an interrupt, it proceeds as follows: after saving the registers of the current user so that the information stored in them will be available when execution resumes on this user's task, the routine reads a byte from each of the input ports. This is done first so that the inputs will always occur at the same time.

Next, the data is output to the terminals. To accomplish this task, a task-status byte is reserved in memory for each user. This byte is a 1 if the terminal is in an output mode, a 2 if the user terminal is in an input mode, and a 0 if the user's task is currently executing without performing I/O operations.

When the I/O has been taken care of, the processor determines which user-task is to be serviced next. The timer-out routine switches current user-tasks, proceeding to work on the task not most recently processed unless that user is still in an input or output mode. If that user is in an I/O mode, control will go back to the task that was being executed when the timer-out interrupt occurred.

This switching process allows both users to "simultaneously" be served by the same processor. At least to human perception, the service appears to be simultaneous. The flowchart in figure 4 supplies a more

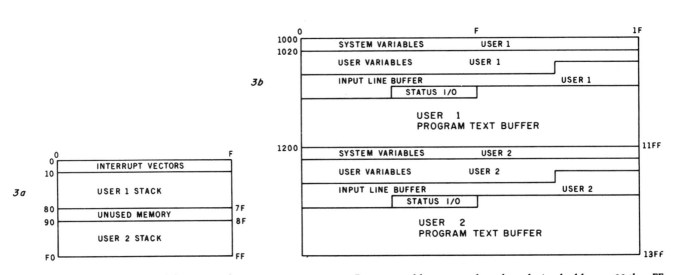

Figure 3: *Map of memory use of the 8088 multi-user operating system. Programmable memory from hexadecimal addresses 00 thru FF is contained in the 8155 integrated circuit and is used chiefly to hold the pushdown stack for each user. Memory from hexadecimal locations 1000 thru 13FF is in the 8185 device, and stores various data belonging to the two user-tasks. Memory from hexadecimal addresses 0F800 to 0FFFF takes the form of EPROM in the 8755A, which stores the operating system.*

detailed accounting of how the multitasking takes place. The assembly code that actually performs the multitasking may be seen in listing 1.

In Conclusion

The hardware discussed in this article is really a bare-bones system. Through the use of more memory (both programmable and read-only memory), as well as through the use of peripheral controllers and pro-grammable interrupt controllers, the whole system could be made to run very efficiently in a multi-user or multiprocessor environment. The possibilities of the new technological developments are impressive.

In the future I will try to let you know about some of the other 16-bit microprocessors. I'd like to wait until I get some evaluation hardware, so that I can relay firsthand experience. ■

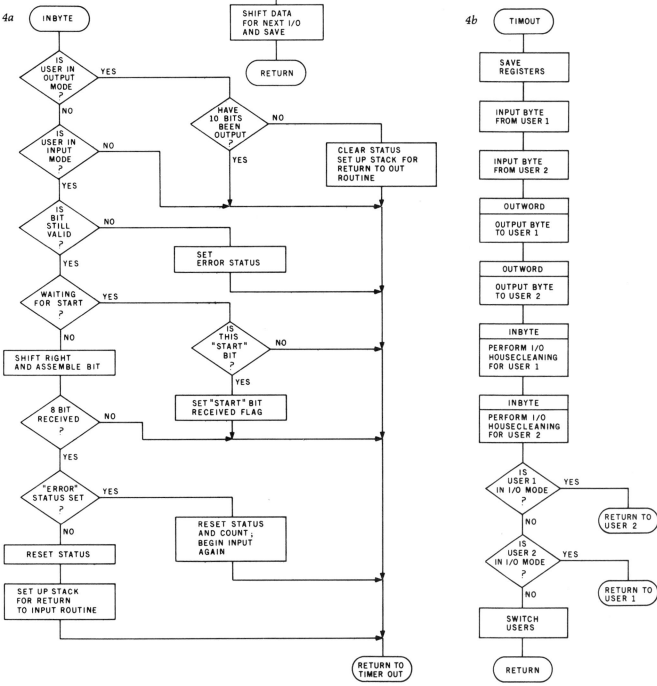

Figure 4: *Flowchart of the multitasking routine of listing 1, which divides the resources of the 8088 system between the two users.*
(4a) Routine to receive input from one of the users.
(4b) Routine to handle timing out of the time-arbitration counter.
(4c) Routine to send output from one of the users.

Listing 1: *Multitasking code that allows two users to be served by the same processor, seemingly simultaneously. Here it is written in assembly language for the 16-bit Intel 8088 microprocessor. When no user requires service, the processor executes a tight loop. When some operation must be carried out, this routine supervises the process. Various I/O operations and counter events cause this code to be entered. The algorithm is shown in flowchart form in figure 4.*

Hexadecimal Address	Hexadecimal Code		Line	Label	Instruction Mnemonic	Operand	Commentary
FE28	EB5990		1174		JMP	USER?	
FE2B	EBFE		1175	IORTI:	JMP	IORTI	;LOOPS TO ITSELF
FE2D	A0AD00	R	1176	CIRT:	MOV	AL,BYTEIN	;RETURNS HERE
FE30	C3		1177		RET		
			1178				
FE31	50		1179	CO:	PUSH	AX	;SAVE REGISTERS
FE32	D1E0		1180		SAL	AX,1	;SHIFT LEFT TO SET
FE34	0D000F		1181		OR	AX,0F00H	;OR TO SET UP STOP
FE37	A3AB00	R	1182		MOV	WORDOT,AX	
FE3A	C606AA000090	R	1183		MOV	OUTCYC,0	;RESET OUTCYCLES
FE40	B001		1184		MOV	AL,1	;SET STATUS TO OUT
FE42	EBD8		1185		JMP	COMP	;SEE IF NEED TO GO
FE44	58		1186	CORT:	POP	AX	
FE45	C3		1187		RET		
FE46	E81801		1188	TIMEOUT:	CALL	SVREG	;SAVE REGISTERS
FE49	BA0200		1189		MOV	DX,INPORT	
FE4C	EC		1190		IN	AL,DX	;INPUT USER 1
FE4D	8AE0		1191		MOV	AH,AL	
FE4F	BA01F0		1192		MOV	DX,INPRT2	
FE52	EC		1193		IN	AL,DX	;INPUT USER 2
FE53	50		1194		PUSH	AX	;SAVE FOR FUTURE USE
FE54	8BC8		1195		MOV	CX,AX	;INPUT DATA,SAVE
FE56	BA0100		1196		MOV	DX,OUTPORT	;SET UP OUTPUT, USER 1
FE59	E85700		1197		CALL	OUTWORD	;OUTPUT
FE5C	8926A700	R	1198		MOV	STACKP,SP	;NEXT BIT OR STOP BIT
FE60	BA00F0		1199		MOV	DX,OUTPT2	
FE63	B80800		1200		MOV	AX,00008H	
FE66	8ED0		1201		MOV	SS,AX	;SET UP SEGMENTS FOR USER 2
FE68	B82000		1202		MOV	AX,20H	
FE6B	8ED8		1203		MOV	DS,AX	
FE6D	8B26A700	R	1204		MOV	SP,STACKP	
FE71	E83F00		1205		CALL	OUTWORD	;OUTPUT NEXT BIT OR CHIP BIT
FE74	BA01F0		1206		MOV	DX,INPRT2	
FE77	E84E00		1207		CALL	INBYTE	;PROCESS INPUT/OUTPUT CYCLES, USER 2
FE7A	59		1208		POP	CX	;RESTORE INPUTS
FE7B	8ACD		1209		MOV	CL,CH	
FE7D	BA0200		1210		MOV	DX,INPORT	;PROCESS INPUT/OUTPUT ,USER2
FE80	E84500		1211		CALL	INBYTE	
FE83	A0AE00	R	1212	USER?:	MOV	AL,STATUS	;DETERMINE WHICH USER TO RESTORE
FE86	2403		1213		AND	AL,03H	
FE88	7406		1214		JZ	CKU2	;IF UI IN CO OR CI
FE8A	B80800		1215		MOV	AX,00008H	;UI IN CO OR CI
FE8D	EB0F90		1216		JMP	PRETI	;GO TO U2
FE90	A0AE02	R	1217	CKU2:	MOV	AL,STATS2	
FE93	2403		1218		AND	AL,03H	
FE95	7509		1219		JNZ	PRET	;USER 2 IN CO OR CI,RETURN TO UI
FE97	36A10C00		1220	SWUS:	MOV	AX,SS:STACKS	
FE9B	350800		1221		XOR	AX,0008H	;SWITCH USER FROM PREVIOUS
FE9E	8ED0		1222	PRETI:	MOV	SS,AX	
FEA0	D1E0		1223		SAL	AX,1	
FEA2	D1E0		1224		MOV	SP,STACKP	
FEA	8ED8		1225		SAL	A,1	
FEA6	8B26A700	R	1226		MOV	DS,AX	;SET UP STOCK & DATA SEG
FEAA	07		1227		POP	ES	
FEAB	5D		1228		POP	BP	;RESTORE REGISTERS
FEAC	5F		1229		POP	DI	
FEAD	5E		1230		POP	SI	
FEAE	5A		1231		POP	DX	
FEAF	59		1232		POP	CX	
FEB0	5B		1233		POP	BX	
FEB1	58		1234		POP	AX	

FEB2	CF		1235		IRET		;RETURN TO PLACE WHERE INTERRUPTED
FEB3	A1AB00	R	1236	OUTWORD:	MOV	AX,WORDDOT	;LOAD WORD OUT
FEB6	8A1EAE00	R	1237		MOV	BL,STATUS	;LOAD STATUS BYTE
FEBA	80CBFE		1238		OR	BL,0FEH	
FEBD	F6D3		1239		NOT	BL	;MAKE BL = 00 IF IN, CO OR 01 IF NOT
FEBF	0AC3		1240		OR	AL,BL	
FEC1	EE		1241		OUT	DX,AL	;OUTPUT BYTE ; OUTPUT BYTE
FEC2	D1F8		1242		SAR	AX,1	;SHIFT FOR NEXT BIT
FEC4	A3AB00	R	1243		MOV	WORDDOT,AX	;AND SAVE WORD FOR NEXT CYCLE
FEC7	C3		1244		RET		
FEC8	8A1EAE00	R	1245	INBYTE:	MOV	BL,STATUS	
FECC	8AFB		1246		MOV	BH,BL	;SEE IF USER IN OUTPUT MODE
FECE	80E301		1247		AND	BL,01H	
FED1	7431		1248		JZ	CKIN	;NO, GO TO CKIN
FED3	FE06AA00	R	1249		INC	OUTCYC	;IN OUTPUT MODE, INCREMENT BITS OUT
FED7	803EAA000A	R	1250		CMP	OUTCYC,10	;OUTPUT 10 BITS?
FEDC	7515		1251		JNE	BRET	;NO RETURN
FEDE	C606AE000090	R	1252		MOV	STATUS,00H	;YES, RESET STATUS AND
FEE4	BB44FE	R	1253		MOV	BX,OFFSET (CGROUP:CORT)	;SET UP RETURN
FEE7	8926A700	R	1254	RSST:	MOV	STACKP,SP	;FOR CHARGE-OUT OR CHARGE-IN
FEEB	83C414		1255		ADD	SP,20	
FEEE	53		1256		PUSH	BX	
FEEF	8B26A700	R	1257		MOV	SP,STACKP	
FEF3	59		1258	BRET:	POP	CX	
FEF4	8926A700	R	1259		MOV	STACKP,SP	;SET UP REGISTERS FOR USER 1
FEF8	33C0		1260		XOR	AX,AX	
FEFA	8ED0		1261		MOV	SS,AX	
FEFC	8ED8		1262		MOV	DS,AX	
FEFE	8B26A700	R	1263		MOV	SP,STACKP	
FF02	51		1264		PUSH	CX	
FF03	C3		1265		RET		
FF04	8ADF		1266	CKIN:	MOV	BL,BH	;SEE IF IN INPUT MODE
FF06	80E302		1267		AND	BL,02H	;IF NOT, RETURN (THRU BRET)
FF09	74E8		1268		JZ	BRET	
FF0B	EC		1269		IN	AL,DX	;INPUT AGAIN & VERIFY VALID DATE
FF0C	3AC1		1270		CMP	AL,CL	;VALID,
FF0E	7408		1271		JZ	BITGD	;YES, BIT IS GOOD
FF10	800EAE008090	R	1272		OR	STATUS,80H	;NO, BIT "ERROR" IN STATUS
FF16	EBDB		1273		JMP	BRET	
FF18	80E180		1274	BITGD:	AND	CL,80H	;WAITING FOR START BIT?
FF1B	80E704		1275		AND	BH,04H	
FF1E	7435		1276		JZ	WAITST	;YES, GO TO WAITST
FF20	A0AD00	R	1277		MOV	AL,BYTEIN	;GET BYTE SO FAR
FF23	D0E8		1278		SHR	AL,1	;SHIFT ONCE FOR NEW BIT
FF25	0AC1		1279		OR	AL,CL	
FF27	A2AD00	R	1280		MOV	BYTEIN,AL	;SAVE BYTE IN
FF2A	FE06A900	R	1281		INC	INCYCL	
FF2E	803EA90008	R	1282		CMP	INCYCL,8	;SEE IF 8 BITS IN
FF33	75BE		1283		JNE	BRET	;--NO
FF35	C606A9000090	R	1284		MOV	INCYCL,0	;YES RESET COUNT OF BITS IN
FF3B	A0AE00	R	1285		MOV	AL,STATUS	
FF3E	2480		1286		AND	AL,80H	;SEE IF BAD BIT IN MIDDLE
FF40	7408		1287		JZ	DELFI	;--NO, CHARACTER GOOD,
FF42	C606AE000290	R	1288		MOV	STATUS,2	;BAD UNIT, RESET STATUS AND
FF48	EBA9		1289		JMP	BRET	
FF4A	C606AE000090	R	1290	DELFI:	MOV	STATUS,0	;RESET STATUS
FF50	BB2DFE	R	1291		MOV	BX,OFFSET (CGROUP:CIRT)	
FF53	EB92		1292		JMP	RSST	;PREPARE FOR RETURN
FF55	0AC9		1293	WAITST:	OR	CL,CL	;SEE IF START BIT IN
FF57	759A		1294		JNZ	BRET	;--NOT YET
FF59	C606AE000690	R	1295		MOV	STATUS,06H	;--YES, RESET STATUS
FF5F	EB92		1296		JMP	BRET	
FF61	891E0000	R	1297	SVREG:	MOV	BL1,BX	
FF65	5B		1298		POP	BX	;SAVE REGISTERS
FF66	50		1299		PUSH	AX	

FF67	FF360000	R	1300		PUSH	BL1	
FF6B	51		1301		PUSH	CX	
FF6C	52		1302		PUSH	DX	
FF6D	56		1303		PUSH	SI	
FF6E	57		1304		PUSH	DI	
FF6F	55		1305		PUSH	BP	
FF70	06		1306		PUSH	ES	
FF71	8926A700	R	1307		MOV	STACKP,SP	;AND SET STACK & DATA
FF75	8CD1		1308		MOV	CX,SS	
FF77	33C0		1309		XOR	AX,AX	;SEGMENTS FOR USER 1.
FF79	8ED0		1310		MOV	SS,AX	
FF7B	8ED8		1311		MOV	DS,AX	
FF7D	8B26A700	R	1312		MOV	SP,STACKP	
FF81	36890E0C00		1313		MOV	SS:STACKS,CX	
FF86	53		1314		PUSH	BX	
FF87	C3		1315		RET		
– – – –			1316	CODE ENDS			
– – – –			1317	CONST2 SEGMENT			
FFF0			1318	ORG 0FFF0H			
FFF0	B8 – – – –	R	1319		MOV	AX,DGROUP	
FFF3	8ED8		1320		MOV	DS,AX	
FFF5	BC7F0090	R	1321		MOV	SP,OFFSET(STK)	
FFF9	EA		1322	DB	0EAH		BOOTSTRAP
FFFA	94FD	R	1323	DW OFFSET INIT			
FFFC	0000		1324	DW 0			
– – – –			1325	CONST2 ENDS			
			1326				
			1327				
			1328				
– – – –			1329	CONST	SEGMENT		
			1330				
0055	483F		1331	HOW	DB	'H?',CR	
0057	0D						
0058	4F4B		1332	OK	DB	'OK',CR	
005A	0D						
005B	573F		1333	WHAT	DB	'W?',CR	
005D	0D						
005E	53		1334	SORRY	DB	'S',CR	
005F	0D						

Listing 2: *A benchmark program in Tiny BASIC that can be used to compare execution speeds of various computer systems. It is used here to test the efficiency of the multitasking system.*

```
100  REM 8088 TINY BASIC BENCHMARK
110  REM SINGLE USER-300 BPS
120  LET A = 0
130  PRINT"START"
140  FOR B = 0 TO 25
150  FOR X = 0 TO 1000
160  LET A = A + 1
170  NEXT X
180  LET A = 0
190  NEXT B
200  PRINT"DONE"
210  END
```

I/O Expansion for the Radio Shack TRS-80

Part 1: Principles of Parallel Ports

I receive a lot of mail: enough that I'm beginning to feel like the "Dear Abby" of the personal computer ranks. The sources of the letters range from high school students asking for advice on science fair projects to major corporations seeking consultant services. Even though it takes considerable time to answer this mail, I regard it as a significant opportunity to gauge reader interest. Every letter in some way contributes to my choice of article topics, either through suggestions or by continued occurrence of similar questions.

Recently, my mail has been dominated by owners of the Radio Shack TRS-80 Model I, thirsting for hardware expansion by means other than Tandy Corporation equipment. The majority of questions concern connection of my interfaces to the TRS-80 expansion connector.

In general, I have tried to present projects that are computer independent. That is, the interfaces described are driven through parallel input/output (I/O) ports rather than directly from a computer bus. This had not been a problem in the past, because virtually all of the early personal computers incorporated some parallel I/O capability. For those experimenters interested in enhanced I/O capabilities, I presented the article

"Memory-Mapped I/O" in the November 1977 *BYTE* on page 10 (reprinted in *Ciarcia's Circuit Cellar* Volume I, BYTE Books), which detailed parallel-port construction.

In the 2½ years since that article was first published, a number of

---

A *port* is a hardware channel for the computer to transmit and to receive *data* via an *external* peripheral device.

---

significant changes have occurred in personal computing. Most importantly, the Radio Shack TRS-80, the Apple II and the Commodore PET were introduced. The difficulty in maintaining and operating a computer is no longer a serious consideration for most computer enthusiasts. Much of my mail indicates that a new explanation of parallel and serial I/O is in order, and that it is time for hardware-expansion circuits to be detailed.

This is the first of a two-part article on serial and parallel I/O port expansion of the TRS-80. The first part emphasizes parallel I/O, and the second

part is concerned with serial interfacing. The result will be a complete Radio Shack software-compatible communications interface capable of supporting a variety of serial- and parallel-interfaced peripheral devices. The hardware was designed and the components were selected to be economical to build and easy to check out. First, here is a brief review of the basics.

What Is an I/O Port?

Just as some people are initially confused with the terms *hardware* and *software*, some find the concept of input and output ports difficult to understand without substantial explanation. The classical definition: a *port* is a hardware channel for the computer to transmit and receive *data* via an *external* peripheral device. The key words in this definition are *external* and *data* which imply externally collected information; the channel through which this data is obtained is called a *port*. A printer is a typical external peripheral device. The characters to be printed are sent from the computer to the printer. In some of the more sophisticated units, status signals such as *busy* and *out of paper* are returned to the computer from the printer.

Ports can be either parallel or serial. In *parallel* mode, data is transferred in increments equivalent to the word size of the computer. On the Z80, for instance, an 8-bit microprocessor, an output instruction through a parallel port transfers 8 bits at a time. A 16-bit processor such as the Intel 8086 transfers data in 16-bit increments. The number of bits transmitted simultaneously by a parallel port is dependent upon the size of the microprocessor data bus and how many bits the processor can transfer simultaneously.

However, serial data is always transmitted a single bit at a time, according to a fixed schedule defined by the data rate (usually expressed in bits per second, or *bps*) and a few specific options. The microprocessor has no single instruction that transmits serial data. It must rely on another device called a universal asynchronous receiver/transmitter (UART) to put the data word into serial form and transmit it. Any communication between the processor and the UART is in parallel form and is done through the processor's memory reference or I/O data-transfer instructions. A more in-depth discussion of serial ports will be presented in Part 2 of this article.

Address, Data, and Control Buses

Consider a computer system that includes a printer, video terminal with keyboard, and an audio cassette recorder as peripherals. Data would have to be relayed to the printer, to and from the video terminal, and to and from the cassette recorder. How can the computer tell the difference between data destined for the terminal and the data destined for the printer?

Most microprocessors incorporate a bidirectional *data bus* and an *address bus*: this is shown in figure 1. To keep track of the data transfer between the processor and its peripherals, the system uses a quantity of control signals which together can be called the *control bus*. The usual 8-bit processor has an 8-bit data bus, a 16-bit address bus, and a dozen or so control signals.

When the microprocessor is reading a data byte from memory, the address of the memory location being referenced is placed on the address bus. Memory information stored at that location goes on the data bus and flows from memory to the processor. When data is being written into memory, the operation is reversed. A 16-bit address bus allows the processor to directly address 65,532 (ie: 64 K) memory locations.

In an 8080 or Z80 processor there is a specific set of instructions that perform input/output functions. The operation of these I/O instructions is similar to that of memory-reference instructions, except that only 8 bits of the address bus are used. These 8 bits

Photo 1: *There are a variety of ways to decode the address for a particular input/output (I/O) port from the signals present on the address bus. The least expensive method uses inverters and printed-circuit-board jumpers to select the correct logic polarities. Three address lines are connected through each 7404 hex inverter with two possible connections for each address line. A connection to the upper trace on the circuit board decodes a logic 1; a connection to the lower trace decodes a logic 0.*

Photo 2: *A more expensive and more easily changed addressing scheme employs dual-in-line-pin (DIP) switches and exclusive-NOR gates. The schematic diagram for this is shown in figure 3b.*

designate one of 256 possible I/O ports. In the case of the example system, a separate port address would be used for each peripheral.

Keeping track of bus direction and information flow is a matter of properly decoding the control signals during program execution. In a Z80 for instance, any memory-reference operation is signified by the control signal $\overline{MREQ}$ in the processor going to a logic 0, or low, state. An input or output operation is designated by the $\overline{I/OREQ}$ control signal being at logic 0.

The direction of the data bus depends on whether the processor is trying to read or to write data. If the processor is in a read mode, the $\overline{RD}$ control signal becomes a logic 0; if the processor is writing, the $\overline{WR}$ line is in the 0 state. Monitoring these four lines, $\overline{MREQ}$, $\overline{I/OREQ}$, $\overline{RD}$, and $\overline{WR}$, gives us all the information necessary to support I/O decoding functions. Figure 2 demonstrates how these control outputs are combined for system use.

Address Decoding

So far we have discussed how to determine when the processor wants to send a character to an output device. In such an operation the $\overline{I/OREQ}$ and $\overline{WR}$ lines are both low. To tell the difference between data for the printer and data for the terminal, we must decode the 8-bit port address.

The port address is determined by the logic voltages present on the low-order eight lines (that is, the 8 least-significant bits) of the address bus during I/O operations. Various techniques can be employed to decode these lines. Figure 3 outlines a few simple methods. The objective, whatever the logic employed, is to produce a single pulse (ie: a strobe) whenever the logic states representing a particular address appear on the address bus. To eliminate false outputs when the processor is executing instructions not dealing with I/O, it is best to combine control and address signals as demonstrated in figure 4.

If you own a 6800- or 6502-based system, you have probably noticed that the processor has no special I/O instructions. This does not mean that these processors have no external communications capability, only that these processors communicate with

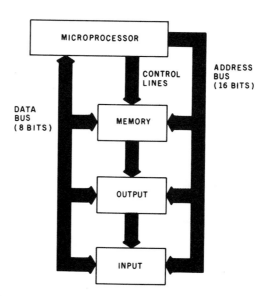

Figure 1: *Block diagram of a microcomputer system that uses an 8-bit microprocessor such as the Z80. This system uses bussing techniques that are both multiplexed and bidirectional.*

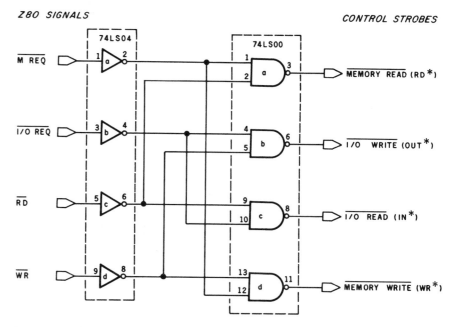

Figure 2: *Control signals on the Z80 microprocessor. The Z80 uses a variety of control signals to keep data flowing at the right time and in the right direction. Four control signals are as follows: the $\overline{MREQ}$ line goes to a low state (ie: a logic 0) when a memory-reference operation is in progress; the $\overline{I/OREQ}$ line goes to a low state when an input/output (I/O) operation is in progress; the $\overline{RD}$ line goes low when the processor is reading data from memory or from a peripheral device; the $\overline{WR}$ line goes low when the processor is writing data to memory or to a peripheral device. The $\overline{RD}$ and $\overline{WR}$ signals control the direction that data flows along the bidirectional data bus. Monitoring these four lines gives us all the information necessary to support I/O decoding functions.*

Signals from the four processor control lines are logically combined to form control-strobe signals that perform specific functions. The characters in parentheses give the names by which the control-strobe signals are known in the documentation for the Radio Shack TRS-80.

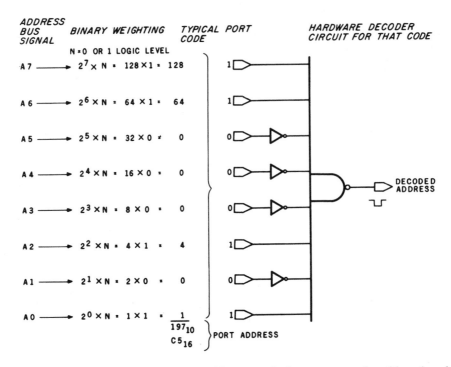

Figure 3a: *Various methods can be employed to decode the address signals that appear on the address bus during I/O operations. Here, various inverters and an eight-input NAND gate are hardwired in a configuration that will produce a logic 0 output for one of 256 possible I/O port addresses. The logic 0 output can be used to activate the interface for the peripheral device. Here the circuit decodes the address hexadecimal C5, or decimal 197.*

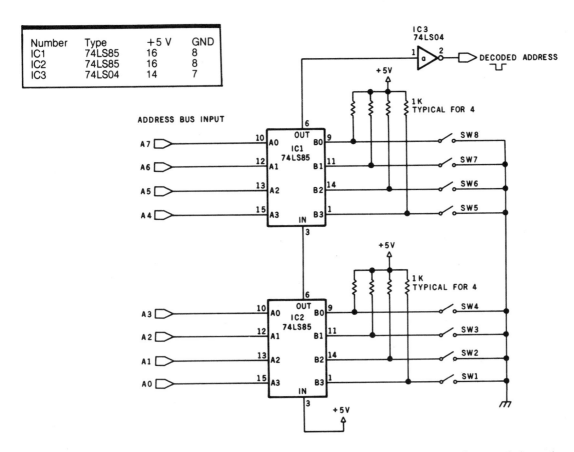

Figure 3b: *Another method of decoding an address signal. Two 4-bit comparators can be cascaded together to decode an 8-bit address. The desired 8-bit port address is set up on switches SW8 thru SW1. When the combination of high and low logic states that corresponds to the desired address appears on the address bus, the output signal produced at pin 2 of IC3 (the 74LS04 inverter) will go low to a logic 0 state. This decoding method allows the port addresses to be easily changed, but the method here is considerably more expensive than the decoding method shown in figure 3a. The switches are single-pole, single-throw (SPST) types; an open switch shows logic 1, and a closed switch shows logic 0.*

peripheral devices differently. How can we discover this different method? Let us begin by looking closely at the I/O functions of the 8080 and Z80 that we have just discussed.

A close inspection of the I/O functions of an 8080 or Z80 should point out that the I/O instructions bear a surprising resemblance to memory-reference instructions. The 6800 and 6502 microprocessors actually allocate a certain portion of their memory address space to be decoded and to function as I/O ports.

This technique, which can be used on the Z80 and 8080 just as easily, has certain advantages in speed and ease of use over direct I/O instructions. This technique is referred to as *memory-mapped I/O*. An illustration of the logic associated with this method is in figure 5. For a more rigorous analysis of memory-mapped I/O, I refer you to the November 1977 "Ciarcia's Circuit Cellar" article

previously mentioned.

The final area for consideration is the actual transfer of data to and from the bidirectional data bus. The circuits of figure 4 and figure 5 tell only *when* the I/O operation occurs. Additional logic has to be provided to place data on the bus during an input instruction or to latch and hold the contents of the data bus during output instructions.

When the 8080 or Z80 assembly-language instruction OUT (N),D is executed, the contents of the accumulator, D, are placed on the data bus and written into device N. The same is true for the BASIC-language instruction OUT N,D. The data is actually valid during only a few clock cycles, perhaps 500 ns. Making this data available for longer periods of time requires the addition of an 8-bit *latch*: the latch is made from a set of clocked flip-flops.

The output lines are attached to the data bus. When the proper output instruction is executed, signified by a strobe signal from our address and I/O WRITE decoder circuit as shown in figure 6, the contents of the data bus are transferred into the 8-bit register in synchronization with the processor clock signal. This combination of circuitry is commonly called an *8-bit latched parallel output port*.

External devices cannot be directly connected to the data bus for input, because of the possibility that interference and bus-loading problems will result. A three-state buffer is used as a *gate* to allow signals from the peripheral device to be placed onto the bus at the appropriate time.

During an input operation the process used for output is reversed. When the proper input sequence is executed, signified by the appropriate output from the address decoder and I/O READ decoder, the 8-bit three-state buffer is strobed into operation during the few clock cycles it takes for the processor to execute the input instruction. Logic levels present on the buffer input lines during that instant become impressed onto the data bus and are transferred into the accumulator. Figure 6 shows the logic elements that perform these functions.

Add Parallel I/O to Your TRS-80

I have been told that the TRS-80 Model I is currently the largest-selling personal computer. Unfortunately it

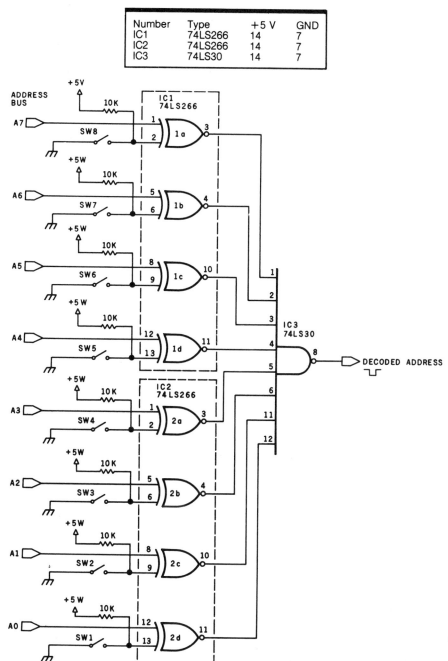

Number	Type	+5 V	GND
IC1	74LS266	14	7
IC2	74LS266	14	7
IC3	74LS30	14	7

Figure 3c: *Another method of decoding an 8-bit address, using exclusive-NOR gates and an eight-input NAND gate. As in figure 3b, the desired port address is set up on switches SW8 thru SW1.*

is not configured to be easily interfaced to the projects I present each month. The widely sold Level II BASIC, 16 K-byte memory version has no parallel I/O capability, aside from the single-bit cassette-motor control. With the addition of the expansion interface, the user gets one parallel output port and one half (ie: 4 bits) of an input port. If these ports are used, as Radio Shack intends, to drive a printer, then the only way to provide usable parallel I/O capability is to add a separate I/O interface.

Considering the pertinent elements of the previous discussions, it is easy to construct both parallel input and parallel output ports for the TRS-80. The interface shown in figure 7 provides one input and one output port. The signals necessary to drive this interface are available on the forty-pin expansion connector of the keyboard/processor unit or on connector J2 on the expansion interface. In either case, a separate +5 V supply is necessary to power the circuit. The signals on the expansion connector are listed in table 1, and the pinouts are shown in figure 8.

The schematic diagram of figure 7 shows a port address FF. To set another port address simply refer to figure 3 and 4 and place the switches for the proper code.

There are many other methods for implementing I/O capability. An 8255 programmable peripheral interface, a parallel I/O device, could have been used. The circuit I have chosen to present is intended to be inexpensive and easy to operate. By minimizing potential parts-acquisition problems and keeping down the software handshaking necessary when using large-scale circuits like the 8255, I hope to enable many TRS-80 owners to build the circuit and use it to attach other "Circuit Cellar" projects to their computer system.

Those experimenters who hesitate to build hardware might want to purchase the entire communications interface. An assembled and tested unit, with power supply and containing a parallel port (for the Centronics printer) and a serial RS-232C-compatible interface, is available. The complete communications unit, called the COMM-80, will be presented in part 2 of this article and

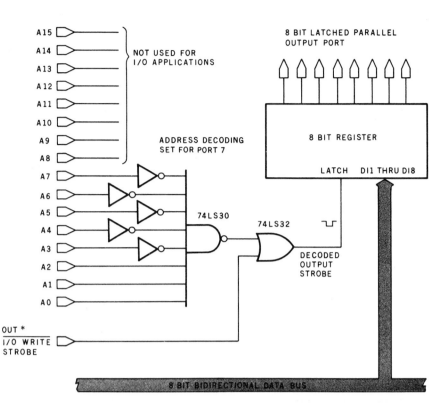

Figure 4a: *Block diagram of a typical parallel output port. The logic that decodes the 8-bit port address was shown in three forms in figure 3. The signal from the address-decoding circuit is logically combined with one of the control signals from figure 2 (I/O WRITE) to produce an output strobe signal that activates the 8-bit output latch register.*

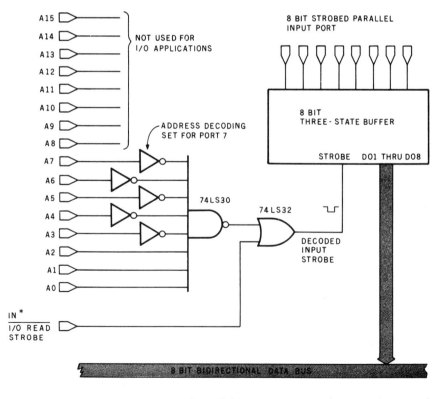

Figure 4b: *Block diagram of a typical parallel input port. Note the resemblance to the output port of figure 4a.*

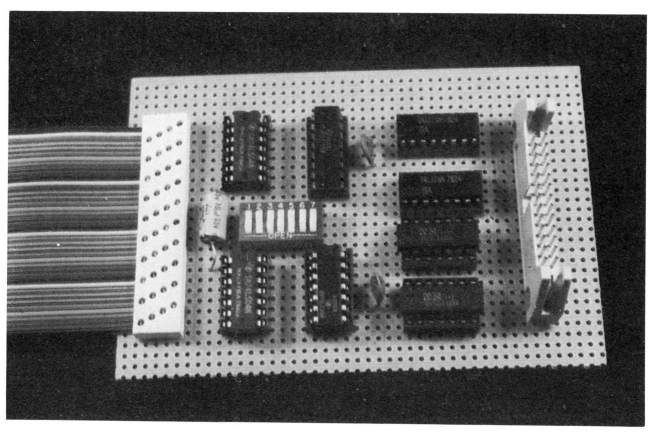

Photo 3: *Prototype of an 8-bit I/O port for the Radio Shack TRS-80. The ribbon cable at left connects to the expansion port on the keyboard/processor unit. The two I/O ports are brought out to the ribbon-cable connector on the right edge of the board.*

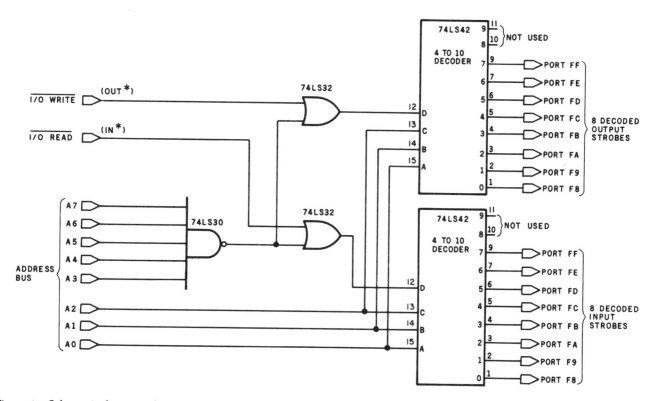

Figure 4c: *Schematic diagram of a circuit that produces eight decoded input-strobe signals and eight decoded output-strobe signals. The port addresses produced are hexadecimal F8 thru FF.*

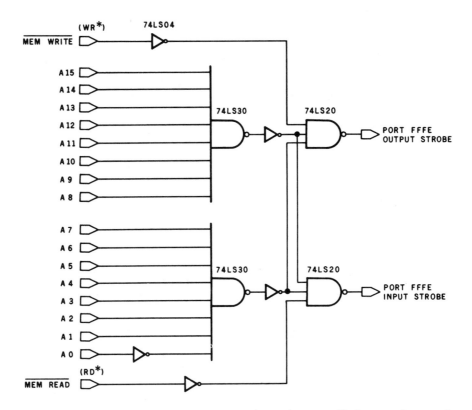

Figure 5: *Memory-mapped input and output. Some microprocessors do not have specific input and output instructions. In systems that use such microprocessors, the I/O port hardware is wired as a memory location; I/O operations take place using the memory-reference instructions (eg: load-into-accumulator and store-in-memory instructions) of the microprocessor. This type of addressing is called memory-mapped I/O, and all sixteen lines on the address bus must be decoded to perform an I/O operation.*

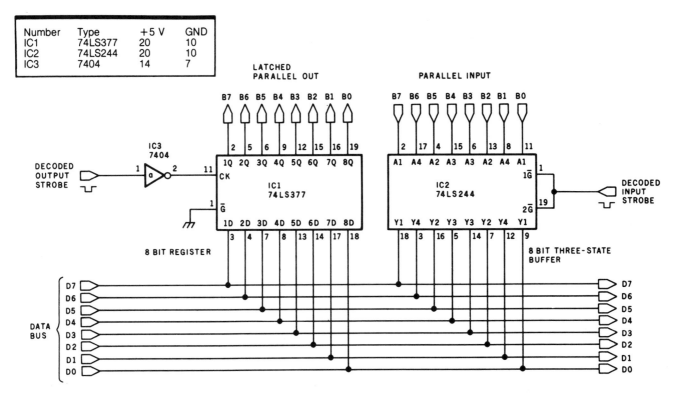

Figure 6: *Data connections in input and output ports. Once the proper port address has been decoded and combined with the read- or write-control signal to form an I/O strobe signal, the actual process of accessing the data bus for data transfer is relatively easy.*

For input to the accumulator (the most common pathway for I/O), a three-state buffer is used in conjunction with the decoded input-strobe signal that controls the enable line of the buffer.

For output from the accumulator, an 8-bit latch is connected to the data bus. During the execution of the output instruction, the contents of the data bus are clocked into the latch register and are latched there by the output-strobe signal.

Number	Type	+5 V	GND
IC1	74LS125	14	7
IC2	74LS125	14	7
IC3	74LS75	5	12
IC4	74LS75	5	12
IC5	74LS155	16	8
IC6	74LS04	14	7
IC7	74LS04	14	7
IC8	74LS30	14	7

is available for $179.95 from:

MicroMint Inc
917 Midway
Woodmere NY 11598
Telephone (516) 374-6793

(New York residents please add applicable sales tax.)■

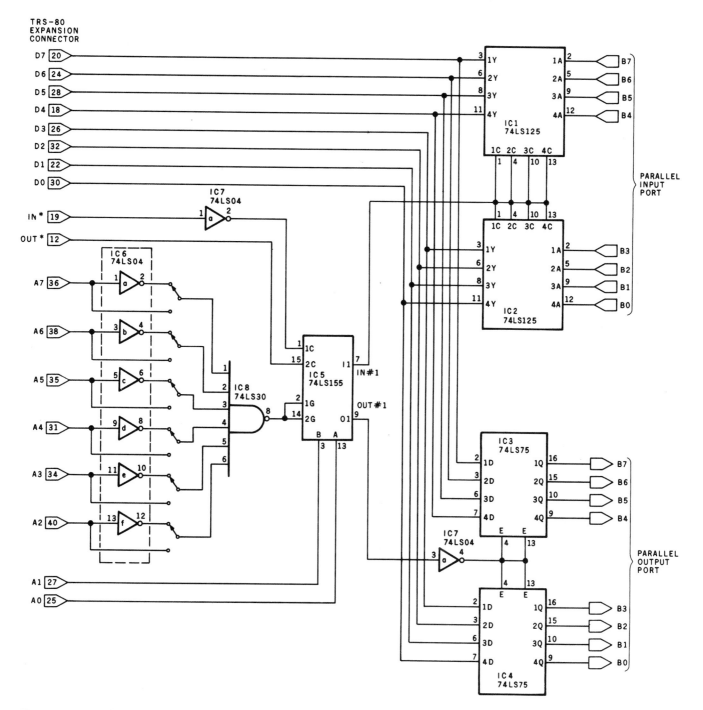

Figure 7: *A complete, economical, parallel I/O interface circuit for use with the Radio Shack TRS-80 computer, or with other computers that use a similar bidirectional data bus. This interface can be connected directly to the expansion connector at the rear of the TRS-80 keyboard/processor unit, or it can be connected through the expansion-interface unit. As the circuit is shown here, there are six presently undefined additional strobes available on IC5. These six strobes can be used to support three additional ports. Refer to figure 3 and figure 4 to determine the proper selection of the I/O port address for this interface.*

Pin Number	Signal Name	Description
1	RAS*	row-address strobe output for 16-pin dynamic memories
2	SYSRES*	system-reset output, low during power-up initialization or when the reset switch is depresssed
3	CAS*	column-address strobe output for 16-pin dynamic memories
4	A10	address output
5	A12	address output
6	A13	address output
7	A15	address output
8	GND	signal ground
9	A11	address output
10	A14	address output
11	A8	address output
12	OUT*	peripheral-write strobe output
13	WR*	memory-write strobe output
14	INTAK*	interrupt-acknowledge output
15	RD*	memory-read strobe output
16	MUX	multiplexer control output for 16-pin dynamic memories
17	A9	address output
18	D4	bidirectional data bus
19	IN*	peripheral-read strobe output
20	D7	bidirectional data bus
21	INT*	interrupt input (maskable)
22	D1	bidirectional data bus
23	TEST*	placing a logic 0 on this line causes a high-impedance condition on address lines A0 thru A15, data lines D0 thru D7, WR*, RD*, IN*, OUT*, RAS*, CAS*, and MUX
24	D6	bidirectional data bus
25	A0	address output
26	D3	bidirectional data bus
27	A1	address output
28	D5	bidirectional data bus
29	GND	signal ground
30	D0	bidirectional data bus
31	A4	address bus
32	D2	bidirectional data bus
33	WAIT*	processor-wait input, to allow for slow memory
34	A3	address output
35	A5	address output
36	A7	address output
37	GND	signal ground
38	A6	address output
39	—	on Level I machines: low-current +5 V output; on Level II machines: no connection
40	A2	address output

Table 1: *Description of function for the pins on the expansion port at the rear of the TRS-80 keyboard/processor unit. This pin assignment is also used in expansion slots in the expansion-interface unit. This information is provided through the courtesy of Radio Shack, a division of Tandy Corporation.*

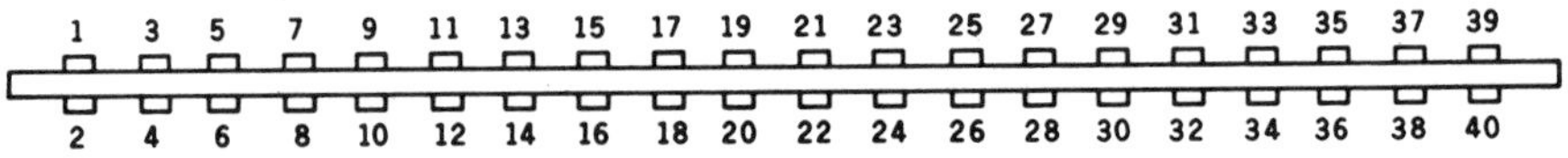

Figure 8: *The configuration of output pins on the expansion port on the rear of the TRS-80 keyboard/processor unit. See table 1 for an explanation of the function of each pin.*

QUESTIONS, QUESTIONS, QUESTIONS

Dear Steve,

I have a couple of questions regarding your article "I/O Expansion for the Radio Shack TRS-80, Part 1." It appears that figure 7 is a diagram of the prototype board pictured in photo 3. Where do the capacitors come in? And what are their values?

I know just enough about electronics to get myself into trouble. know *what* the components are and *how* they work, but I don't know how to match them up into a working circuit.

Also, could you furnish more information about using the extra logic on IC5 to operate the three additional ports? I am particularly interested in a combination security system and external-device control and monitor. I don't think 8 bits is enough for what I have in mind.

I have done some figuring on the additional ports. It appears to me that, for each additional port, I will need one 74LS04, one 74LS30, and one 14-pin DIP switch to decode the port address. For input capabilities, I would need two 74LS125s and two 74LS75s.

Since there are four inverters unused on IC7, three could be used with the latches for the three other ports.

Kerry A Wilson

You are correct. Figure 7 is the circuit of photo 3. The extra capacitors are for decoupling and protective filtering. These components are added because they are a good idea and not because they are necessary for the port function described. Whenever TTL (transistor-transistor logic) components are used in a design, capacitors are attached across the power-supply pins to eliminate noise in the power wiring. The value is usually 0.01 μF to 0.1 μF, and one should be added for every three integrated circuits (this figure is variable and depends on circuit density and power consumption as

well).

The larger capacitor is a 10 μF electrolytic type which is attached between +5 V and ground where the power enters the board. Whenever an interface is remotely powered, it is possible that the wires attaching it to the power source will pick up noise. Adding a capacitor at the end of the power cable helps reduce this noise. The exact value is a function of cable impedances and circuit reactance, but, in low-current circuits, 10 μF to 100 μF is acceptable. High-quality designs may be a little more particular, and tantalum electrolytics are generally used.

The additional logic necessary to expand figure 7 for three more ports would be six 74LS125s, six 74LS75s, and three of the remaining inverter sections of IC7. For each port, you would duplicate the circuit of ICs 1, 2, 3, 4, and 7a; however, use the other strobe lines on IC5, the 74LS155. Those lines are described in detail in the second part of my article. The addressing for the other ports is already decoded in the original circuit. As the switches are shown, the first port is 00. The other three will be 01, 02, and 03 respectively.

Be careful to keep your wiring short and neat because this circuit is attached to the main computer bus. If the computer malfunctions, then you may need to add extra buffers to the data and I/O buses.
...Steve

I/O Expansion for the TRS-80
Part 2: Serial Ports

In Part 1, I discussed the attachment of parallel input and output ports to the Radio Shack TRS-80 computer. This was basically a response to the many inquiries I have had on TRS-80 interfacing. As usual, it was a general presentation, intended to first enlighten the reader with interfacing concepts and then tender a few alternative circuits for construction. While TRS-80 owners benefit most directly, many computers have similar bus structure and can just as easily accommodate parallel input/output (I/O) expansion.

The presentation of a serial interface for the TRS-80 required a little more thought. Parallel ports are strictly hardware devices which in their simplest form only require execution of a single assembly-language or BASIC instruction to function efficiently. A serial interface, on the other hand, needs a software program to direct its operation. The many registers and buffers involved in the serial communication process must be synchronized by the execution of a serial-driver routine stored in memory. Any design for a serial port has to take into account the capabilities and memory location of this routine. even the most splendid hardware circuit would be a failure if the software driver interfered with other computer functions.

To eliminate any potential problems that might occur, I decided to make my design completely software-compatible with existing TRS-80 serial-driver routines. This does not necessarily minimize circuit complexity by any means, but it greatly enhances potential user acceptance.

I was equally concerned with the power requirements and physical configuration. Radio Shack sells a

This RS-232C interface design is compatible with existing TRS-80 serial-interface control software.

serial-interface board for the TRS-80, but it cannot be operated independently and requires integral attachment to the expansion interface

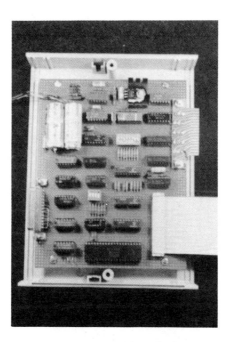

Photo 1: *Prototype of the COMM-80 interface. The ribbon cable at the lower right connects to the expansion-bus port (either the expansion connector on the keyboard/processor unit or connector J2 on the expansion interface). The edge connector at the upper right is for the Centronics-compatible, parallel printer port. The RS-232C DB-25S connector is at the lower left.*

module. The expansion interface and one serial port add $400 to the cost of the basic computer. Also, with its present hardwired addressing, the TRS-80 can support only one serial port and one parallel printer port.

Depending upon the intended application, you may not need the extra functions (eg: disk controller and memory expansion) provided in the expansion interface. The $300 outlay for the expansion interface is an extraordinary expense if you merely intend to attach a modem and use the TRS-80 as a terminal on a timesharing network, such as the Source or MicroNet. Rather than duplicate what I consider to be a restrictive hardware configuration, I have attempted to present a cost-effective communications interface that gives more flexibility in use and has a better price/performance ratio.

The COMM-80 Communications Interface

The approach I decided to take was to combine elements from Part 1 of this article with this one, and produce a stand-alone serial/parallel interface which could plug directly into the expansion-bus connector (the keyboard-unit expansion connector or connector J2 on the expansion interface). Designated the COMM-80, the unit includes a 50 to 19,200 bit per second (bps) RS-232C serial port, a full 8-bit-in/8-bit-out parallel printer port, an auxiliary expansion-port edge connector, and switch-selectable addressing which allows a single TRS-80 to simultaneously connect up to sixteen COMM-80 interfaces. A block diagram of the COMM-80 is presented in figure 1, and a picture of the prototype is in photo 1.

What Is a Serial Port?

Communication between computers, terminals, and other peripheral devices can be in either serial or parallel mode. In parallel mode, the entire information segment (ie: data word) is transmitted or received simultaneously in a single time frame. In serial mode, this same information is divided into its constituent bits and these bits are transmitted individually over a longer period of time. In cases where high-speed data rates are involved, such as in interaction with a floppy-disk drive, the communication is usually in parallel and can involve as many as forty data and control lines. Serial mode is generally used for lower-speed exchanges.

Photo 2: *Here are two ways of adding RS-232 communication capability to the Radio Shack TRS-80. The COMM-80 unit is shown on the left; the combination of the Radio Shack expansion interface and serial-interface board is shown on the right.*

Photo 3: *A TRS-80 equipped with Level II BASIC, the COMM-80 interface, and a Novation CAT modem can be used as a remote terminal for a time-sharing service such as the Source.*

An example a little closer to home is the addition of a video terminal and a printer to a computer system. Both the terminal and printer are designed to accept American Standard Code for Information Interchange (ASCII) coding, which requires only 7 bits to define a character.

The connections between the computer and the video terminal can be either serial or parallel. The choice in this case is not determined by data rate but by expense. Parallel communication is relatively easy and inexpensive for a computer. Few components are involved, and a 6-foot length of nine-conductor cable (seven lines to carry the 7-bit ASCII data, one line each for data strobe and ground) will not cost too much. Serial interfacing is another matter entirely.

Microprocessors do not naturally communicate in serial format. There are no single machine-language instructions to perform this function. To serialize data we must add a separate hardware device called a universal asynchronous receiver/transmitter (UART). It looks just like a parallel port to the processor, but internally the UART is a very complicated device.

A UART is a special large-scale integration (LSI) circuit that accepts a data byte in parallel form from the processor and converts it into a universally accepted serial format. Any two terminals set at the same data-transmission rate could conceivably be interconnected to communicate, regardless of internal operating-system differences. The expense for this flexibility is in the neighborhood of $200 to $500 per data channel, depending upon the computer bus configuration.

Transmitting Serial Data

Serial data can be transmitted in either *synchronous* or *asynchronous* format. I will address this discussion only to the latter format since asynchronous communication is the technique employed in the COMM-80. The asynchronous format allows unlimited time gaps to occur between transmission of characters.

The internal structure of a UART consists of a separate parallel-to-serial transmitter and a serial-to-parallel receiver joined by common programming pins. The two sections can be used independently provided they adhere to the same bit-format options. Sending a character from the processor is simply a matter of performing a parallel-output operation to the UART. The decoded-output strobe loads the UART with the data and initiates the serialization process.

Figure 2 shows a plot of logic levels versus time during the transmission of a single character. When no data is being sent, the data-transmission line remains in a logic 1 state. A 1-to-0 high-to-low transition on the line signifies that a character is being sent. The first bit is called a *start bit*. The next 5 to 8 bits are data; these are followed by a parity bit. Finally, the end of transmission is defined by the addition of 1 or 2 stop bits at the end of the character. The start, stop, and parity bits are all added as part of the UART's function.

Meanwhile, the receiver section of the UART is continuously monitoring the input line for the start bit of a character. When the start bit comes, the following data bits are placed into a holding register and their parity is checked against the state of the parity bit. Completion is signaled by setting a *data-available* flag. This flag, plus others defining *buffer status, parity,* and *overrun errors,* is read by the processor to determine when input data is ready or when another character can be transmitted. The individual pin functions of a typical UART are described in table 1.

RS-232C Interface Characteristics

So far, I have discussed only serialization of the data. I have said nothing about voltages or logic conventions associated with control of the information transmitted between

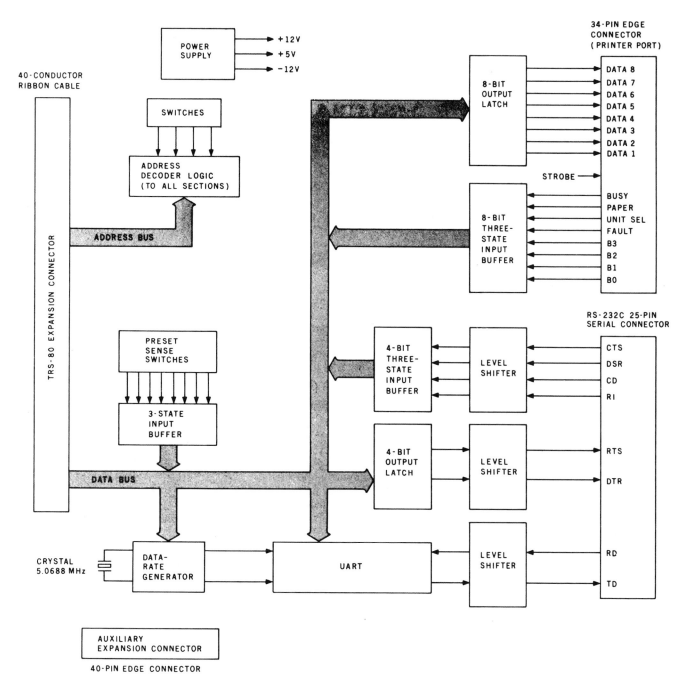

Figure 1: *Block diagram of components and data flow in the COMM-80 serial and parallel interface for the Radio Shack TRS-80.*

equipment. The Electronic Industry Association (EIA) RS-232C electrical specification defines voltage levels and control signals: a logic level 1 is called a "mark" or "off" and is considered to be anything more negative than −3 V. A logic 0 is called a "space" or "on" and is considered to be anything more positive than +3 V. As a rule, designers tend to use +12 V and −12 V for the 0 and 1 logic states.

In addition to standardizing the serial format, the EIA also specifies that the connector for RS-232C be a 25-pin, D subminiature type (called a *DB-25*). The pin assignments and functions are shown in table 2.

The COMM-80 Hardware

The COMM-80 is driven only by signals present on the buses of the computer. All sections communicate with the processor as memory-mapped or directly addressed input/output ports. Figure 3 illustrates the complete schematic diagram of the COMM-80 interface in three sections.

There are two major sections: parallel printer port and serial port. They are joined together by a common address-decoding circuit and power supply.

Address Decoding

A standard TRS-80 expansion interface has an edge connector commonly called the Centronics printer port. It actually combines an 8-bit parallel output port and a 4-bit parallel input port. The addressing for this section is hardwired for hexadecimal memory location 37E8. Part of this same address decoder is used for the Radio Shack serial-interface board. Coincidentally, the Radio Shack serial interface is decoded to use I/O port addresses E8 thru EB for data-transfer and control functions.

The address-decoding section of the COMM-80, consisting of IC1 thru IC7, is designed to decode this set of

Pin Number	Name	Symbol	Function
1	V_{cc} Power Supply	V_{cc}	+5 V Supply
2	V_{GG} Power Supply	V_{GG}	−12 V Supply (Not connected on AY-5-1015
3	Ground	V_{GL}	Ground
4	Received Data Enable	RDE	A logic 0 on the receiver-enable line places the received data onto the output lines.
5			These are the eight data output
6		RD8	lines. Received characters are
7		RD7	right justified; the least-significant
8	Received Data Bits	RD6	bit (LSB) always appears on RD1.
9		RD5	These lines have three-state outputs.
10		RD4	
11		RD3	
12		RD2	
		RD1	
13	Parity Error	PE	This three-state line goes to a logic 1 if the received-character parity does not agree with the selected parity.
14	Framing Error	FE	This three-state line goes to a logic 1 if the received character has no valid stop bit.
15	Over-Run	OR	This three-state line goes to a logic 1 if the previously received character is not read (DAV line not reset) before the present character is transferred to the receiver-holding register.
16	Status Word Enable	SWE	A logic 0 on this three-state line places the status word bits (PE, FE, OP, DAV, TBMT) onto the output lines.
17	Receiver Clock	RCP	This line will contain a clock whose frequency is sixteen times the desired receiver data rate.
18	Reset Data Available	RDAV	A logic 0 will reset the DAV line.
19	Data Available	DAV	This three-state line goes to a logic 1 when an entire character has been received and transferred to the receiver holding register.
20	Serial Input	SI	This line accepts the serial bit input stream. A marking (logic 1) to spacing (logic 0) transition is required for initiation of data reception.
21	External Reset	XR	Resets shift registers. Sets SO, EOC, and TBMT to a logic 1. Resets DAV, and error flags to 0. Clears input data buffer. Must be tied to logic 0 when not in use.
22	Transmitter Buffer Empty	TBMT	The three-state transmitter buffer-empty flag goes to a logic 1 when the data bits holding register may be loaded with another character.

Table 1: *Pin functions for the AY-5-1013, AY-5-1015, or COM2017 UARTs.*

addresses as well as a range of other addresses. The range for the printer port is hexadecimal memory addresses 3708 to 37F8, and the serial range is hexadecimal I/O addresses 08 to F8. Figure 4 illustrates the switch settings for the different ranges.

There is a particular rationale for setting up the addresses this way. A user attaching a COMM-80 to his system would naturally set the switches for the range E8 thru EB, and the interface would then be completely compatible with standard TRS-80 software. Should an expansion-

Figure 2: *Logic levels plotted against time during the transmission of an 8-bit data word in asynchronous serial format.*

Pin Number	Name	Symbol	Function
23	Data Strobe	$\overline{DS}$	A strobe on this line will enter the data bits into the data-bits-holding register. Initial data transmission is initiated by the rising edge of $\overline{DS}$. Data must be stable during entire strobe.
24	End of Character	EOC	This line goes to a logic 1 each time a full character has been transmitted. It remains at this level until the start of transmission of the next character.
25	Serial Output	SO	The entire character is transmitted bit by bit (that is, serially) over this line. It will remain at logic 1 when no data is being transmitted.
26	Data Bit Inputs	TD1	There are up to 8 data-bit-input lines available.
27		TD2	
28		TD3	
29		TD4	
30		TD5	
31		TD6	
32		TD7	
33		TD8	
34	Control Strobe	CS	A logic 1 on this lead will enter the control bits (EPS, NB1, NB2, TSB, NP) into the control-bits-holding register. This line can be strobed or hardwired to a logic 1 level.
35	No Parity	NP	A logic 1 on this lead will eliminate the parity bit from the transmitted and received character (no PE indication). The stop bit(s) will immediately follow the last data bit. If not used, this lead must be tied to a logic 0.
36	Number of Stop Bits	TSB	This lead will select the number of stop bits (1 or 2) to be appended immediately after the parity bit. A logic 0 will insert 2 stop bits.
37	Number of Bits Per Character	NB2	These two leads will be internally decoded to select either 5, 6, 7, or 8 data bits per character.
38		NB1	

NB2	NB1	bits/character
0	0	5
0	1	6
1	0	7
1	1	8

Pin Number	Name	Symbol	Function
39	Odd/Even Parity Select	EPS	The logic level on this pin selects the type of parity which will be appended immediately after the data bits. It also determines the parity that will be checked by the receiver. A logic 0 will insert odd parity, and a logic 1 will insert even parity.
40	Transmitter Clock	TCP	This line will contain a clock whose frequency is sixteen times the desired transmitter data rate.

interface module be added to the system later, the user would merely flip a switch specified by table 3 to change the port address (the expansion interface is set only for 37E8). The switch circuit is shown in figure 4. The system could then accommodate two printers. As table 3 shows, there are sixteen possibilities, so there could be sixteen printers and sixteen serial ports. From this point on, however, I will refer only to the addressing range of E8 thru EB.

The Printer Port Is a Full 8 Bits

Since I explained parallel ports in detail in the last article, I will discuss the printer port briefly. Initially my intention was to provide a general-purpose I/O port so that the user could connect some of my other projects and interface designs. As it worked out, however, I decided to combine efforts and configure the parallel port to serve as the printer port as well. The major difference is that the COMM-80 incorporates a full 8-bit input and a full 8-bit output port. Its address is nominally hexadecimal 37E8 in memory-address space. Writing to memory location 37E8 latches data onto IC14 and IC15 (both 74LS75 devices), and reading memory location 37E8 gates the

printer status signals through the three-state buffer IC19 (a 74LS244 device).

Serial Port

The serial-port section requires four input and four output strobes to operate. As previously mentioned, the serial-port control addresses are nominally set for hexadecimal E8 thru EB. Figure 5 more explicitly illustrates the hardware derivation of these signals and lists their functions. These strobe signals coordinate the RS-232C handshaking, the sense switches, the data-rate generator, and the UART. All four subsections can be independently controlled in software by reading and writing to the appropriate port address.

The sense switches, for instance, are merely a convenience. It is a way for the user to present a frequently used combination of options. These switches, outlined in figure 6, allow selection of data rate, word length, parity condition, and number of stop bits. There is, however, no physical connection between these switches and the other sections. The software-driver routine coordinates the option selection.

First the routine determines the state of the switches by reading input port E9. It determines from the setting of switches SW6 thru SW8 what data rate the user wants. The particular code for that rate, selected from table 4, is written to output port E9. The remaining switch settings are written into the UART control register EA. Three bits of this output (b_0 thru b_2) and input port E8 are used for the RS-232C handshaking. The data-rate generator is presented in figure 7.

The sense switches are not absolutely necessary for operation of the serial interface. Most software drivers, such as the ST80 program written by Lance Micklus, offer a selection of the options through the keyboard. Separate data rates for the

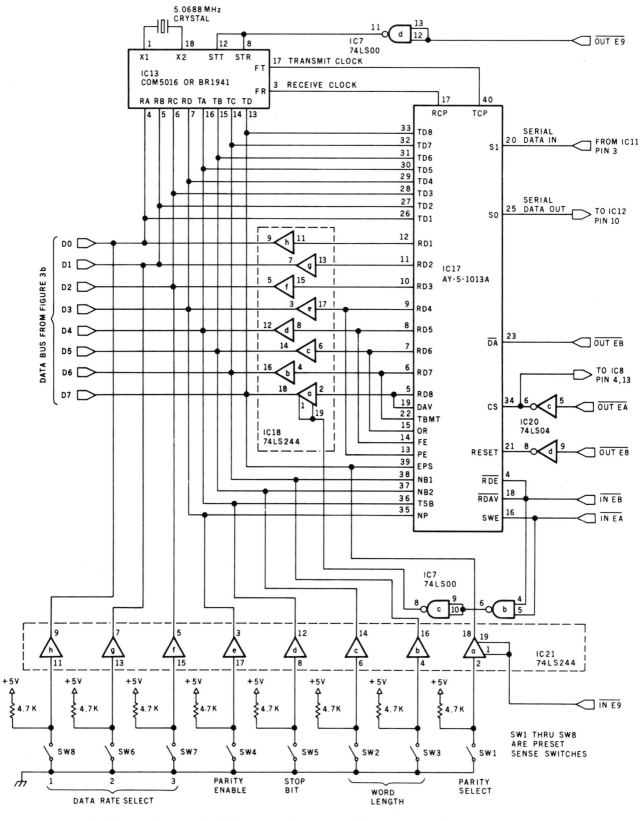

Figure 3a: *Section of schematic diagram of COMM-80 interface circuit. Shown here are the data-rate selector, the UART, and the option-selecting switches. The data-rate selector can be either a COM5016 or a BR1941. Various UARTs can be used instead of the AY-5-1013A, including the TR1602, COM2017, S1883, and TMS6011. A UART that uses a single +5 V power supply, such as the AY-3-1015, may also be substituted.*

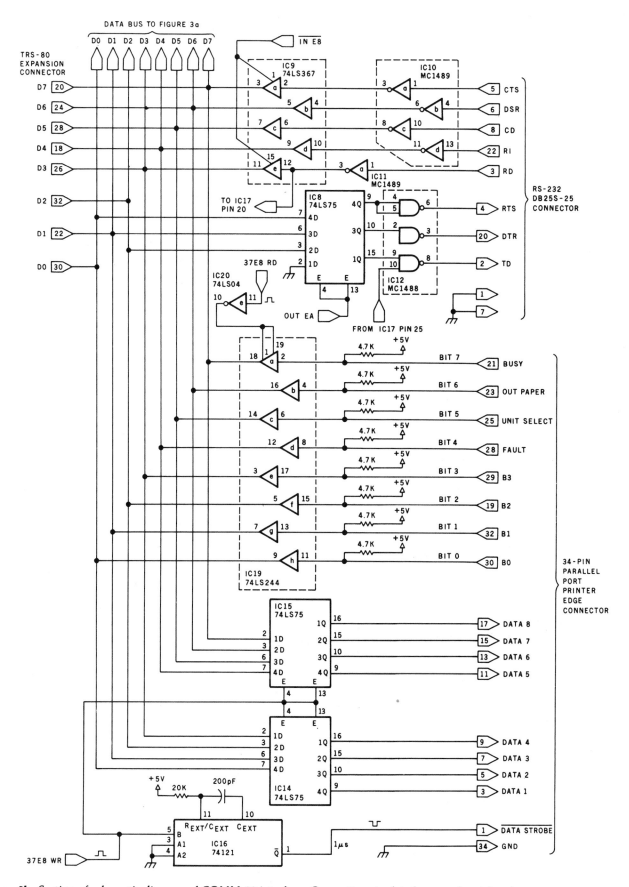

Figure 3b: *Section of schematic diagram of COMM-80 interface. Connections to data buses and peripheral connectors are presented here. Some care must be exercised in connecting the COMM-80 to the expansion bus. It is best to use shielded ribbon cable. The production version of the COMM-80 includes two auxiliary expansion-bus edge connectors, which are like the one on the back of the keyboard/processor unit.*

transmitter and receiver can also be established. This is easily accomplished by a direct output command to the data-rate generator using the codes from figure 6.

From this point on, serial communication proceeds by simply loading the UART with the data to be transmitted (using the Z80 instruction OUT EB) and reading the UART status register to see if the byte has been completely sent or if there is a received data word available (with the IN EA instruction).

The software driver needed for this interface is too long to discuss in this article. Also, since this interface is software-compatible with existing TRS-80 hardware, there is no need to write your own driver routine. There are many sources, including the one listed with this article.

Using the COMM-80

Once you have an RS-232C port installed in your computer, a whole new world of peripherals opens up. The electronics industry has been turning out thousands of printers each year which use the RS-232C interface. For example, if you are interested in word processing, then you can attach a high-quality daisy-wheel printer to your TRS-80. Cer-

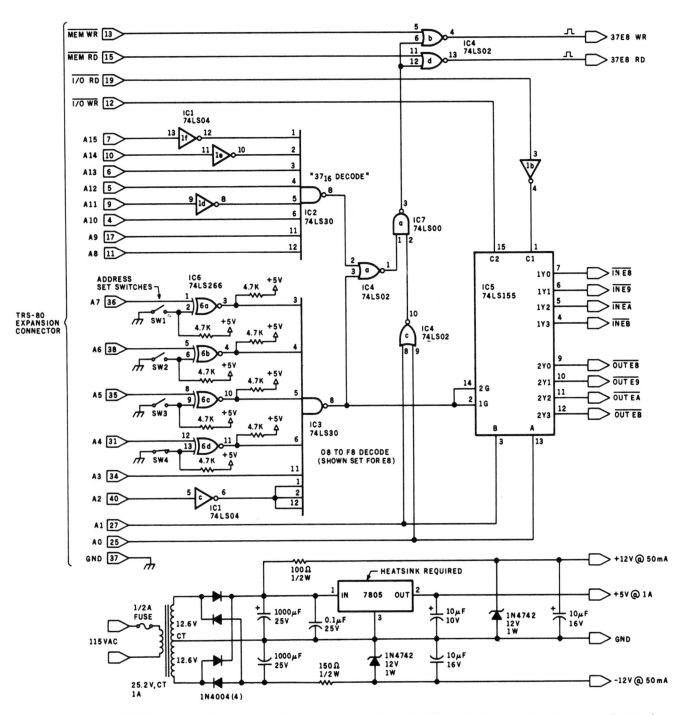

Figure 3c: *Section of COMM-80 interface circuit, including power supply and address-selection circuitry. Power to the interface should not be cut off while the TRS-80 is in operation, lest programs be lost. Both units should be powered up and down simultaneously.*

tain peripherals require a 20 mA current-loop interface; the required circuit is demonstrated in figure 8.

The most obvious application for the COMM-80 is to transform the TRS-80 from a mild-mannered personal computer into a full-fledged computer terminal. Photo 3 shows the system connected to a modem in actual use on the Source timesharing system. Listing 1 is a printout (from an LA36 DECwriter II also connected to the same serial interface) of typical user interaction on this national computer timesharing network. A look at some of the capabilities available through these networks might convince some people to use the network's facilities rather than spend thousands of dollars to build up an independent single-user system. At $2.75 per hour of connect time, it seems a reasonable alternative. For those of you wishing to contact me via the Source, my electronic-mail identification is TCE317. I welcome questions on this or any other topics that I might possibly be able to answer. ∎

The COMM-80 is available assembled and tested in an attractive 20.95 by 15.24 by 6.35 cm (8.25 by 6 by 2.5 inch) enclosure, including expansion-bus connector and cable, an auxiliary 40-pin expansion-port edge connector, a 34-pin Centronics parallel printer port, RS-232C serial port with DB-25S connector, user's manual, power supply, and terminal software.

Price (including shipping): $179.95 (New York residents please add appropriate sales tax).

Also available:
1. Bare bones kit including COM5016, CM2017, printed circuit board and manual, 5.0688 MHz crystal. COMM-80 kit for $55.00
2. TRS-80 1 foot 40 conductor ribbon cable with connectors COMM-80 cable for $20.00
3. TRS-80 terminal software for $20.00

Order from:
The MicroMint Inc
917 Midway
Woodmere, NY 11595
(516) 374-6793

These prices are valid until Dec. 31, 1981. Call for prices after that date.

Pin 1	PGND — Protective Ground This is chassis or equipment ground. It may also be tied to signal ground.
Pin 2	TD — Transmit Data This is the serial data from the terminal to the remote receiving equipment. When no data is being sent it is in a marking (1) condition.
Pin 3	RD — Receive Data This is the serial data from the remote equipment which is transmitted to the terminal.
Pin 4	RTS — Request to Send Controls the direction of data transmission. In full-duplex operation an ''on'' sets transmit mode and an ''off'' sets non-transmit mode. In half-duplex operation an ''on'' inhibits the receive mode and an ''off'' enables it.
Pin 5	CTS — Clear to Send Signal from the modem to the terminal indicating ability to transmit data. An ''on'' is ''Ready'' and an ''off'' is ''not ready.''
Pin 6	DSR — Data Set Ready Signal from the modem to the terminal. An ''on'' condition indicates that the modem is ready.
Pin 7	SGND — Signal Ground
Pin 8	CD — Carrier Detect An ''on'' indicates reception of a carrier from the remote data set; ''off'' indicates no carrier is being received.
Pin 20	DTR — Data Terminal Ready: ''on'' connects the communication equipment to the communications channel; ''off'' disconnects the communications equipment from the communications channel.
Pin 22	RI — Ring Indicator An ''on'' indicates that a ringing signal is being received on the communications channel.

Table 2: *Designations of pins on the DB-25 connector when used for communication with an RS-232C interface system and description of corresponding signals.*

Address Range	SW1	SW2	SW3	SW4
08 thru 0B	Closed	Closed	Closed	Closed
18 thru 1B	Closed	Closed	Closed	Open
28 thru 2B	Closed	Closed	Open	Closed
38 thru 3B	Closed	Closed	Open	Open
48 thru 4B	Closed	Open	Closed	Closed
58 thru 5B	Closed	Open	Closed	Open
68 thru 6B	Closed	Open	Open	Closed
78 thru 7B	Closed	Open	Open	Open
88 thru 8B	Open	Closed	Closed	Closed
98 thru 9B	Open	Closed	Closed	Open
A8 thru AB	Open	Closed	Open	Closed
B8 thru BB	Open	Closed	Open	Open
C8 thru CB	Open	Open	Closed	Closed
D8 thru DB	Open	Open	Closed	Open
E8 thru EB	Open	Open	Open	Closed
F8 thru FB	Open	Open	Open	Open

Table 3: *Use of the switch-selectable address decoder allows the I/O address range to be varied over the range shown here according to the switch positions specified. (See figure 4.) Radio Shack software uses the address range hexadecimal E8 thru EB.*

T_A	T_B or R_B	T_C R_C	T_D R_D	Data Rate	Clock Frequency
R_A					
0	0	0	0	50	800 Hz
1	0	0	0	75	1200 Hz
0	1	0	0	110	1760 Hz
1	1	0	0	134.5	2152 Hz
0	0	1	0	150	2400 Hz
1	0	1	0	300	4800 Hz
0	1	1	0	600	9600 Hz
1	1	1	0	1200	19.2 kHz
0	0	0	1	1800	28.8 kHz
1	0	0	1	200	32.08 kHz
0	1	0	1	2400	38.4 kHz
1	1	0	1	3600	57.6 kHz
0	0	1	1	4800	76.8 kHz
1	0	1	1	7200	115.2 kHz
0	1	1	1	9600	153.6 kHz
1	1	1	1	19200	316.8 kHz

Table 4: *Chart to select data rates for the COM5016 data-rate generator. Transmission and reception rates may be set independently, according to the parameters specified here.*

Listing 1: *Part of the output generated during a timesharing session on the Source, in which the TRS-80 equipped with the COMM-80 and a modem was used as a terminal. The Source is a service of the Source Telecomputing Corporation of McLean, Virginia. The hard copy was produced by an LA36 DECwriter connected to the TRS-80 through the COMM-80.*

```
>DATA SYSCOM

************************ SYSTEM COMMANDS ************************

COMMAND                      DESCRIPTION
-------                      -----------

BASIC          PROGRAM IN THE BASIC LANGUAGE.
CHAT           TALK TO ANOTHER USER ON THE SYSTEM.
CRTLST         DISPLAYS THE CONTENTS OF A FILE, STOPPING EVERY 24
               LINES TO GIVE YOU TIME TO CATCH UP. (TYPING A RETURN
               RESTARTS THE DISPLAY.)
DATA           DISPLAYS CERTAIN TCA LIBRARY PROGRAMS AND DATA BASES.
DATE           GIVES TIME AND DATE.
DEL            DELETES A FILE.
DELAY          AUTOMATICALLY DELAYS OUTPUT TO PRINTING TERMINALS
               WITH SLOWLY RETURNING CARRIAGES.
ED             TEXT EDITOR.
ENTER          TYPE IN A FILE.
FILES          PRINTS THE NAME OF ALL YOUR FILES.
FORTRN         COMPILES A FORTRAN PROGRAM.
ID             SYSTEM SIGN-ON COMMAND.
INFO           DISPLAYS CERTAIN OTHER LIBRARY PROGRAMS AND DATA BASES.
LOAD           LOADS A FORTRAN PROGRAM.

MAIL           INVOKES THE ELECTRONIC MAIL PROGRAM.
NSORT          SORTS A FILE.
OFF            SIGNS A USER OFF THE SYSTEM.
PLAY           PLAYS COMPUTER GAMES.
POST           INVOKES THE CLASSIFIED AD/BULLETIN BOARD PROGRAM.
R              RUNS A LIBRARY PROGRAM.
TIME           DISPLAYS THE TIME USED FOR THE CURRENT SESSION.
RUN            RUNS A LOADED FORTRAN PROGRAM.
TY             LIKE CRTLST, BUT DOES NOT STOP AFTER 24 LINES.
USAGE          SUMMARY OF YOUR SYSTEM USAGE THIS MONTH.

NOTE: A COMPLETE LIST OF SYSTEM DOCUMENTATION AND PROGRAMMING
MANUALS MAY BE VIEWED BY TYPING DATA SYSDOC.
>ONLINE
CL0158    CL0619    TCA056    TCA088    TCA088    TCA290
TCA422    TCA434    TCA516    TCA569    TCA575    TCA612
TCA743    TCA766    TCA830    TCA914    TCB419    TCD011
TCD106    TCD140    TCD202    TCD248    TCD390    TCD419
TCD419    TCD437    TCD444    TCD459    TCD460    TCE052
TCE129    TCE201    TCE217    TCE274    TCE317
>DATA UPI

****************** UNITED PRESS INTERNATIONAL ******************

1) TO ACCESS THE UPI DATANEWS SYSTEM, SIMPLY TYPE 'UPI' AND
PRESS 'RETURN'.
2) THEN SELECT 'NATIONAL', 'REGIONAL' OR 'STATE' NEWS OR 'FEATURES'.
'FEATURES' INCLUDES MOST MAJOR NEWS SYNDICATES (NEW YORK TIMES,
UNITED FEATURES, ETC.) AS WELL AS SYNDICATED COLUMNISTS. FOR A
COMPLETE LIST OF FEATURES, INDEXED BY LOGICAL CONTENT, RETURN TO
THE 'COMMAND' LEVEL, AND TYPE........................UPI
3) SELECT FROM THE 'GENERAL', 'BUSINESS' OR 'SPORTS' CATEGORIES;
 THE SYSTEM WILL THEN ASK YOU FOR ONE OR MORE 'KEYWORDS'.
```

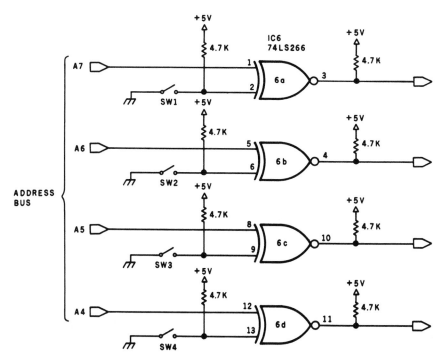

Figure 4: *By closing the proper switches, one of sixteen possible address ranges in the I/O-address space can easily be selected. The switches are optional; the desired address range may be hardwired. For complete compatibility with standard TRS-80 software, the hexadecimal address range E8 thru EB should be chosen.*

Number	Type	+5 V	GND	−12 V	+12 V
IC1	74LS04	14	7		
IC2	74LS30	14	7		
IC3	74LS30	14	7		
IC4	74LS02	14	7		
IC5	74LS155	16	8		
IC6	74LS266	14	7		
IC7	74LS00	14	7		
IC8	74LS75	5	12		
IC9	74LS367	16	8		
IC10	MC1489	14	7		
IC11	MC1489	14	7		
IC12	MC1488		7	1	14
IC13	COM5016	2	11		9
IC14	74LS75	5	12		
IC15	74LS75	5	12		
IC16	74121	14	7		
IC17	AY-5-1013A	1	3	2	
IC18	74LS244	20	10		
IC19	74LS244	20	10		
IC20	74LS04	14	7		
IC21	74LS244	20	10		

Table 5: *Power supplies needed by the integrated circuits in the COMM-80.*

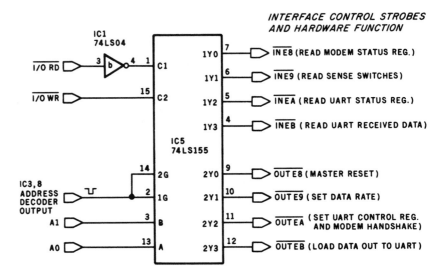

Figure 5: *Detail figure demonstrating interface-control strobes. The address decoder (made up of IC3 and IC6) can be set within the range of hexadecimal 08 to F8. TRS-80 compatibility requires a low address of E8. The output-strobe address notations presented refer only to this setting. Switch settings for other addresses are given in table 3.*

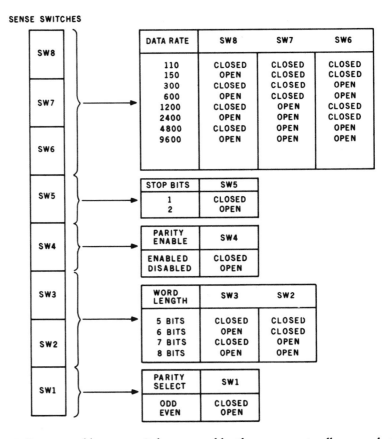

Figure 6: *Programmable sense switches are read by the processor to allow preselection of UART options under program control. The correspondence of options and switches is illustrated here.*

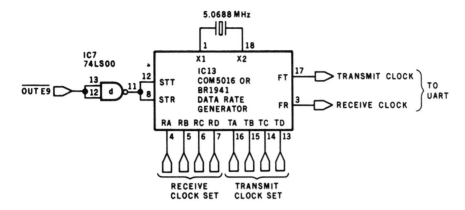

Figure 7: *The data-rate generator determines how fast data is sent and received. Transmission and reception rates can be set independently. The specifications for setting up the various possible data rates on the COM5016 are presented in table 4.*

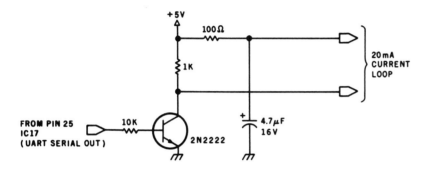

Figure 8: *Some peripheral devices (ie: a Teletype ASR33) must be connected by means of a 20 mA current-loop circuit; such a circuit that can be attached to the COMM-80 is shown here.*

COMM-80

Dear Steve,

I read with great interest your article "I/O Expansion for the TRS-80, Part 2: Serial Ports." However, I have a few questions before I can decide whether or not the COMM-80 is the answer to my problems.

I own a Radio Shack TRS-80, Level II, with 16 K bytes of programmable memory. Eventually I want to expand it to include 32 K bytes, a parallel printer, a telephone modem, and an 8-inch floppy-disk system. Since the Radio Shack floppy-disk system is only a 5-inch disk, I knew I'd eventually need to use equipment from another manufacturer.

Could you please tell me if it is possible to use a disk system with the COMM-80? Would that use the RS-232 port? Could I use both the extra 16 K bytes of memory *and* a disk system at the same time?

Richard L Jamison

The COMM-80 is a serial and parallel port. It can be used with any hardware that will plug into a TRS-80, but it has nothing logically to do with a disk system. However, if you purchase a stand-alone disk controller or memory-expansion module, they would be TRS-BUS driven and could be plugged into the expansion bus connector of the COMM-80.

*It is also my understanding that most disk operating systems for the TRS-80 require a 32 K-byte system to run effectively. The question should be whether you can live without the extra 16 K bytes....**Steve***

Dear Steve,

My colleagues and I have worked in computer-controlled laboratories over the years and are intrigued with teaching and research possibilities with modern microcomputers. The Radio Shack TRS-80 is in some senses a bargain, but as you rightly point out, the I/O (input/output) problem is

serious. Your I/O Expansion articles are suggestive, but questions remain:

● How can the COMM-80 plug into the J2 connector on the Expansion Interface since J2 is the place in the Expansion Interface where it connects itself to the TRS-80? (You must mean the J3 40-pin bus-extension edge connector.)
● How *can* COMM-80 be used with TRS-80 Expansion Interface? How are sixteen units connected to one another, *and* to the TRS-80? Your photo 1 shows only 3 connections: one to TRS-80 keyboard, one for serial printer (RS-232), and one for parallel printer interface.

If sixteen units are indeed all "chainable" to the Expansion Interface and one has a 16-unit bus-extension capacity, we would appreciate literature and specs.

J P Rosenfeld PhD

Sorry about that; you are correct. The signals on J2 and J3 are the same, but the COMM-80 attaches to J3 when using an Expansion Interface. This is also called the screen-printer port in some of Radio Shack's literature.

The COMM-80 is a bus-driven device and can be connected either to the keyboard connector or to J3 as I mentioned. The COMM-80 contains a parallel (Centronics-compatible) and a serial (RS-232C) port. Internally, it has a 4-bit address-selection switch. If you keep the address setting as shipped, the COMM-80 will perform as if it were the serial TRS-232 board installed within the Expansion Interface. However, if you already have a TRS-232 board and want to just add another printer to the system or drive a serial printer in addition to the modem, you would set the COMM-80 to one of the other sixteen addresses. Whether you use two or four COMM-80s, as long as they have different switch settings, they are independent I/O (input/output) channels to the computer.

With a 4-bit address code, up to 16 COMM-80s (or 15 and a TRS-232 board) can conceivably be attached to a TRS-80. Each COMM-80 has an auxiliary TRS-BUS connector that is the same as J3. This means that the second COMM-80 is plugged into the auxiliary connector (creating a daisy chain), since J3 on the expansion interface is already occupied. As a practical matter, however, the cable lengths get to be a problem with over ten units connected together. Some extra buffering might be needed at the TRS-80 end to drive thirty or so feet of bus. I believe Radio Shack sold such a device at one time.

*As far as photo 1 goes, that was my prototype and it did not have the auxiliary expansion-bus connector. The photo on page 62 of June 1980 BYTE is of a production COMM-80 board. The three connectors you see are (clockwise from bottom left) for the auxiliary TRS-BUS, RS-232, and parallel printer. The connections missing in this photo include the 1.5 feet of ribbon cable and the 40-pin expansion-bus edge connector. They would have obscured the picture....***Steve***

INDEX